A Comprehensible Guide to
Servo Motor Sizing

Copperhill Media Corporation
http://www.copperhillmedia.com

A Comprehensible Guide to Servo Motor Sizing

By Wilfried Voss

Published by

Copperhill Media Corporation
158 Log Plain Road
Greenfield, MA 01301

ISBN: 978-0-9765116-1-8

Printed in the United States of America

Limit of Liability/Disclaimer of Warranty

Copperhill Media
http://www.copperhillmedia.com

About this book

After years of developing servo motor sizing programs for Windows, I deemed it necessary to document the motor sizing process beyond the regular help files. This book is the result of this idea.

My bible for motor sizing and the inertia/torque calculation of mechanical components has been for years a faded copy of "The Texonics Motion Cheat Sheet", which was given to me by Mr. Charles Geraldi and for which I will be forever grateful. Years later, I received a similar version, "The Smart Motion Cheat Sheet" by MSI Technologies, Inc., which was created by Brad Grant, P.E. I am sure somebody knows the story of how the motion cheat sheet was originally developed and how it evolved. It is so far the most effective written tool for motor sizing of which I am aware.

Both versions of the motion cheat sheet contain on only a few pages everything I needed to know to create the first version of a motor sizing software under MS-DOS and then under the various Windows versions. The result of all these activities is VisualSizer-Professional™, the most advanced motor sizing software in the business. Many businesses in the motion control industry have chosen to use VisualSizer and have it customized for their purposes.

Through cooperation with many motion control experts, I learned that the servo motor sizing process is a somewhat universal procedure. The calculation of inertia and torque of mechanical components, i.e. the motor load, has not changed since the invention of the electrical motor. All the motor has to do is to match the speed, inertia and torque requirements. However, written works on in-depth motor sizing, besides frequent, but somewhat superficial articles in various motion control publications, are extremely rare.

The motor sizing process involves a number of documented mathematical equations not necessarily with the motor sizing process in mind. This book focuses primarily on servo motor sizing, not on motor technologies and other related specifics. It documents the inertia and torque calculation of standard mechanical components and the motor selection process.

Last, but not least, let me re-iterate a previous statement: The calculation of mechanical loads goes back to the time when Isaac Newton described the three laws of motion. The math of calculating mechanical loads has not changed since then and did not change with the use of electrical motors. It may be that mechanical components became more sophisticated, but, again, that did not change the way we calculate their inertia and torque.

This circumstance may also explain why contemporary motion control literature does not properly document inertia and torque calculations, but seems to focus primarily on modern technology aspects, such as tuning and programming. After all, the purpose of using an electrical motor is to move a mechanical load and the understanding of the mechanical load is as important as programming a duty cycle.

About the author

Wilfried Voss is the President of Copperhill Technologies Corporation, a company specializing in motor sizing software development for various motor manufacturers all over the world. In addition, Copperhill Technologies sells user licenses of its generic motor sizing program, VisualSizer-Professional™. Since 2005, technical literature on all aspects of motion control and fieldbus technologies is included in the product offering.

Mr. Voss has been involved with motion control applications since 1985 as a specialist in the paper industry and in addition, since 1997, is involved with fieldbus technologies, especially CAN (Controller Area Network) related technologies. He has a master's degree in electrical engineering from the University of Wuppertal in Germany. Mr. Voss has traveled the world extensively, settling in New England in 1989. He presently lives in an old farmhouse in Greenfield, Massachusetts with his Irish-American wife and their son Patrick.

Acknowledgements by the author

This book would never have been possible without the vision of Mr. Charles Geraldi, then Sales Manager at Parvex Servo Systems in New York, who in 1992 hired me to develop a motor sizing software for MS-DOS and Windows 3.1. "Not expensive!" he wrote on the two pages from a notepad, which represented the entire outlining of the software. At the time, I had done some motion controller programming, but had never heard of motor sizing.

Another collaborator in this coupe d'état on my person was Edward Crofton, then President of Automation & Servo Technologies, who, in cooperation with Charles Geraldi, was one of my biggest supporters for many years. No one could have imagined that this afternoon meeting in 1992 would lead to creating the most successful software package of its kind. Thanks, guys!
Also thanks to those experts who provided their support and knowledge and helped to make VisualSizer a success: Both John M's of Baldor Motor and Drives, i.e. John Malinowksi and John

Mazurkiewicz, for their kind support and for showing me the best BBQ place in the world (a little shed in Oklahoma).

Thanks also to Paul Derstine of GE Fanuc, a vigorous tester of my program, Craig Ludwick and Nick Johantgen of Oriental Motors U.S.A., Uwe Krauter of Siemens Energy & Automation, Meng King of AC Tech (Lenze) and George Gulalo, President of Motion Tech Trends, who provided ingenious insights to the motor sizing process. Last, but not least, I appreciate the help of Steve Huard of Parker Hannifin, not only a top expert in servo motor technologies, but also top in programming motion control software.

Table of Contents

Overview

The vast majority of automated manufacturing systems involve the use of sophisticated motion control systems that, besides mechanical components, incorporate electrical components such as servo motors, amplifiers and controllers.

The first straightforward task for the motion system design engineer, before tuning and programming the electrical components, is to specify – preferably the smallest - motor and drive combination that can provide the torque, speed and acceleration as required by the mechanical set up.

However, all too often engineers are familiar with the electrical details, but do lack the knowledge of how to calculate the torque requirements of the driven mechanical components. In other cases, they try to size their application around the motor and spend valuable time to figure out how to move the load under the given circumstances. Such an approach will lead to improperly sized motion control applications. The impact, economically as well as technically, will be one of the topics in the following chapters.

Modern motor sizing software packages, such as VisualSizer-Professional™, provide the convenience of computing all necessary equations and selecting the optimum motor/drive combination within minutes; they are, however, mainly used to circumvent the timely process of selecting a motor manually. While motor sizing programs can have an educational value to some degree, most of them do not provide any reference on how the equations were derived.

Some basic knowledge of inertia and torque calculations can have a profound impact on the motion system performance. Simple details, like when to use a gearbox in a motion system, may not only help to reduce costs, but will most certainly improve the system performance.

The following chapters will provide a comprehensive insight into the motor sizing process including detailed descriptions of inertia and torque calculations of standard mechanical components.

Chapter

2

The Importance of Servo Motor Sizing

The importance of servo motor sizing should not be underestimated. Proper motor sizing will not only result in significant cost savings by saving energy, reducing purchasing and operating costs, reducing downtime, etc.; it also helps the engineer to design better motion control systems.

2.1   Why Motor Sizing?

The servo motor represents the most influential cost factor in the motion control system design, not only during the purchasing process, but especially during operation. A high-torque motor will require a stronger and thus more expensive amplifier than smaller motors. The combination of higher torque motor plus amplifier results not only in higher initial expenses, but will also lead to higher operational costs, in particular increased energy consumption. It is estimated, that the purchase price represents only about 2% of the total life cycle costs; about 96% is electricity.

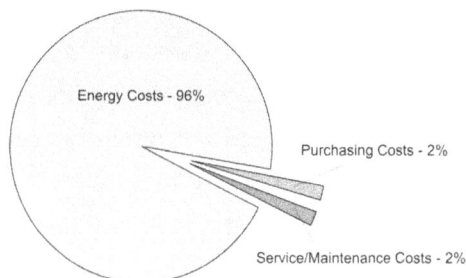

Energy Costs - 96%

Purchasing Costs - 2%

Service/Maintenance Costs - 2%

Picture 2.1.1: Lifecycle costs of an Electrical Motor

Proper servo motor sizing will not only assure best system performance; it also provides considerable cost savings.

The conventional method of servo motor sizing is based on calculations of the system load, which determines the required size of a motor. Standard praxis demands to add a safety factor to the torque requirements in order to cover for additional friction forces that might occur due to the aging of mechanical components. However, the determination of the system load and the selection of the right servo motor can be extremely time consuming. Each motor has its individual rotor inertia, which contributes to the system load torque, since Torque equals Inertia times Acceleration. The calculation of the system torque must be repeated for each motor that is being considered for the application.

As a result, it is not an easy task to select the optimum motor for the application considering the vast amount of available servo motors in the marketplace. Many motors, that are currently in action, have been chosen mostly due to the fact that they are larger than required and were available short-term (e.g. from inventory). The U.S. Department of Energy estimates that about 80% of all motors in the United States are oversized.

The main reasons to oversize a motor are:
- Uncertain load requirements
- Allowance for load increase (e.g. due to aging mechanical components)
- Availability (e.g. inventory)

Not only is the power consumption higher than it should be; there are also some serious technical considerations.

2.2 Technical Aspects

Oversizing a motor is naturally more common than undersizing. An undersized motor will consequently not be able to move the load adequately (or not at all) and, in extreme cases, may overheat and burn out, especially when it can't dissipate waste heat fast enough. Larger motors will stay cool, but if they are too large, they will waste energy during inefficient operation. After all, the motor sizing process can also be seen as an energy balancing act.

AC motors tend to run hot when they are loaded too heavily or too lightly. Servo motors, either undersized or oversized, will inevitably start to vibrate or encounter stalling problems.

One of the major misconceptions during the motion design process is that selecting a larger motor than required is only a small price to pay for the capability to handle the required load, especially since the load may increase during the lifetime of the application due to increased mechanical wear. However, as demonstrated in the picture below, the motor efficiency deteriorates quickly when the motor operates below the designed load.

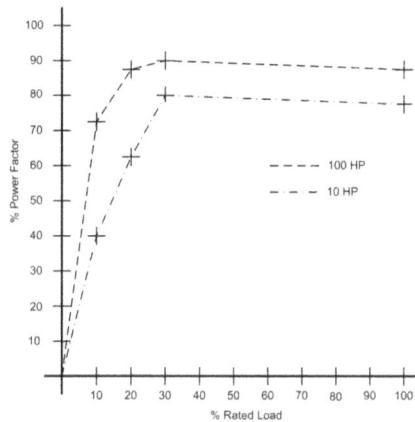

Picture 2.2.1: Example Efficiency vs. Load

Picture 2.2.1 shows an example of two motors, 10 HP and 100 HP. In both cases, there is a sharp decline of the motors' efficiency at around 30% of the rated load. However, the curves as shown in the picture will vary substantially from motor to motor and it is difficult to say when exactly a motor is oversized. As a general rule of thumb, when a motor operates at 40% or less of its rated load, it is a good candidate for downsizing, especially in cases where the load does not vary very much. Servo motor applications usually require short-term operation at higher loads, especially during acceleration and deceleration, which makes it necessary to look at the average (RMS) torque and the peak torque of an application.

There are, however, advantages to oversizing:

➢ Mechanical components (e.g. couplings, ball bearings, etc.) may, depending on the environment and quality of service, encounter wear and as a result may produce higher friction forces. Friction forces contribute to the constant torque of a mechanical set up.

➢ Oversizing may provide additional capacity for future expansions and may eliminate the need to replace the motor.

➢ Oversized motors can accommodate unanticipated high loads.

➢ Oversized motors are more likely to start and operate in undervoltage conditions.

➢ In general, a modest oversizing of up to 20% is absolutely acceptable.

> ➤ High efficiency motors, compared to standard motors, will maintain their efficiency level over a broader range of loads (see picture 2.2.2) and are more suitable for oversizing.[1]

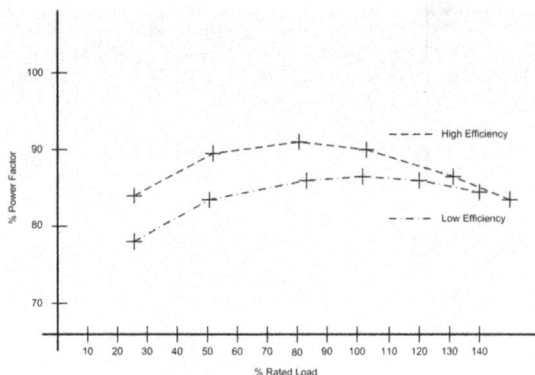

Picture 2.2.2: Example High/Low Efficiency Motors

2.3 The Objective of Motor Sizing

The main objective of motor sizing is based on the good old American sense for business: Get the best performance for the lowest price. As we have learned from a previous chapter, the lifecycle costs of an electrical motor are:

> ➤ Purchasing Costs – 2%
> ➤ Repair, Service, Maintenance, etc. – 2%
> ➤ Operating Costs (Electricity) – 96%

In order to get the best performance for the best price it is mandatory to find the smallest motor that fulfills the requirements, i.e. the motor that matches the required torque as close as possible. The basic assumption (which is true for the majority of cases) is that small torque is in direct proportion to smaller size, lower costs and lower power consumption. Smaller power consumptions also result in smaller drive/amplifier size and price.

From a technical standpoint it is also desirable to find a motor whose rotor inertia matches the inertia of the mechanical setup as close as possible, i.e. the optimum ratio between load to

---

[1] A detailed discussion of the various motor technologies is not in the scope of this book. For further information on high efficiency motors and their selection refer to http://www.motorsmatter.org.

rotor inertia of 1:1. The inertia match will provide the best performance. However, for servo motors a ratio of up to 6:1 still provides a reasonable performance. Any higher ratios will result in instabilities of the system and will eventually lead to total malfunction.

In many cases, it makes sense to add a gear between motor and the actual load. A gear lowers the inertia that is reflected to the motor in direct proportion of the transmission ratio. This scenario allows running smaller motors, however, with the price of the gear added to the system. On the other hand, the price reduction by using a smaller motor/drive combination may more than just compensate for the gear's price.

In review, the objective of motor sizing is to:

> Get the best performance for the best price
> Match the motor's torque with the load torque as close as possible
> Match the motor's inertia with the load inertia as close as possible
> Find a motor that matches or exceeds the required speed

The Motor Sizing and Selection Process

The motor sizing and selection process is based on the calculation of torque and inertia imposed by the mechanical set up plus the speed and acceleration required by the application. The selected motor must be able to safely drive the mechanical set up by providing sufficient torque and velocity. Once the requirements have been established, it is easy to look either at the torque vs. speed curves or motor specs and choose the right motor.

The sizing process involves the following steps:

1. Establishment of motion objectives
2. Selection of mechanical components
3. Definition of a load (duty) cycle
4. Load calculation
5. Motor selection

1. Establishment of motion objectives

- ➢ A written outlining of the motion control application will help to establish the necessary parameters needed for the next steps.
- ➢ Required positioning accuracy?
- ➢ Required position repeatability?
- ➢ Required velocity accuracy?
- ➢ Linear or rotary application?
- ➢ If linear application: Horizontal or vertical application?
- ➢ Thermal considerations – Ambient temperature?
- ➢ What motor technologies are best suited for the application?

2. Selection of mechanical components

The engineer must decide which mechanical components are required for the application. For instance, a linear application may require a leadscrew or a conveyor. For speed transmission a gear or a belt drive may be used.

> ➤ Direct Drive?
> ➤ Special application or standard mechanical devices?
>
> ➤ If linear application: Use of linear motor or leadscrew, conveyor, etc.?
> ➤ Reducer required – Gearbox, belt drive, etc.?
> ➤ Check shaft dimensions – select couplings
> ➤ Mechanical components for speed and acceleration limitations

3. Definition of a load (duty) cycle

The engineer must define the maximum velocity, maximum acceleration, duty cycle time, acceleration and deceleration ramps, dwell time, etc., specific to the application.

> ➤ Define critical move parameters such as velocity, acceleration rate
> ➤ Triangular, trapezoidal or other motion profile?
> ➤ If linear application: Make sure the duty cycle does not exceed the travel range of linear motion device.
> ➤ Jerk Limitation required?
> ➤ Consideration of thrust load?
> ➤ Does the load change during the duty cycle?
> ➤ Holding brake applied during zero velocity?

4. Load calculation

The torque that is required to drive the mechanical set up defines the load. The inertia "reflected" from the mechanical set up to the motor and the acceleration at the motor shaft determines the amount of torque.

> ➤ Calculate inertia of all moving components
> ➤ Determine inertia reflected to motor
> ➤ Determine velocity, acceleration at motor shaft
> ➤ Calculate acceleration torque at motor shaft
> ➤ Determine non-inertial forces such as gravity, friction, pre-load forces, etc.
> ➤ Calculate constant torque at motor shaft
> ➤ Calculate total acceleration and RMS (continuous over duty cycle) torque at motor shaft

5. Motor Selection

The motor must be able to provide the torque required by the mechanical set up plus the torque inflicted by its own rotor. Each motor has its specific rotor inertia, which contributes to

10

the torque of the entire motion system. When selecting a motor the engineer needs to recalculate the load torque for each individual motor.

- ➢ Decide the motor technology to use (DC brush, DC brushless, stepper, etc.)
- ➢ Select a motor/drive combination
- ➢ Does motor support the required maximum velocity? If no, select next motor/drive.
- ➢ Use rotor inertia to calculate system (motor plus mechanical components) acceleration (peak) and RMS torque
- ➢ Does motor's rated torque support the system's RMS torque? If no, select next motor/drive.
- ➢ Does motor's intermittent torque support the system's peak torque? If no, select next motor/drive.
- ➢ Does the motor's performance curve (torque over speed) support the torque and speed requirements? If no, select next motor/drive.
- ➢ If the ratio of load over rotor inertia exceeds a certain range (for servo motors 6:1), consider the use of a gearbox or increase the transmission ratio of the existing gearbox. Servo motors should not be operated over a ratio of 10:1.

note A modest oversizing of the motor of up to 20% is absolutely acceptable. The oversizing factor should be implemented during the torque requirement checks. In this case, it also acceptable to use a higher factor for the acceleration (peak) torque.

Hear Ye! Hear Ye! The motor selection process as described also explains the popularity of motor sizing programs. The process of recalculating the torque requirements for each individual motor/drive combination can be extremely time-consuming considering the vast amount of motors available in the industry. The goal of motor sizing is to find the optimum motor for the application and that can only be accomplished with sufficient choices available, i.e. with a great number of applicable motors.

The following flow chart demonstrates the motor sizing and selection process:

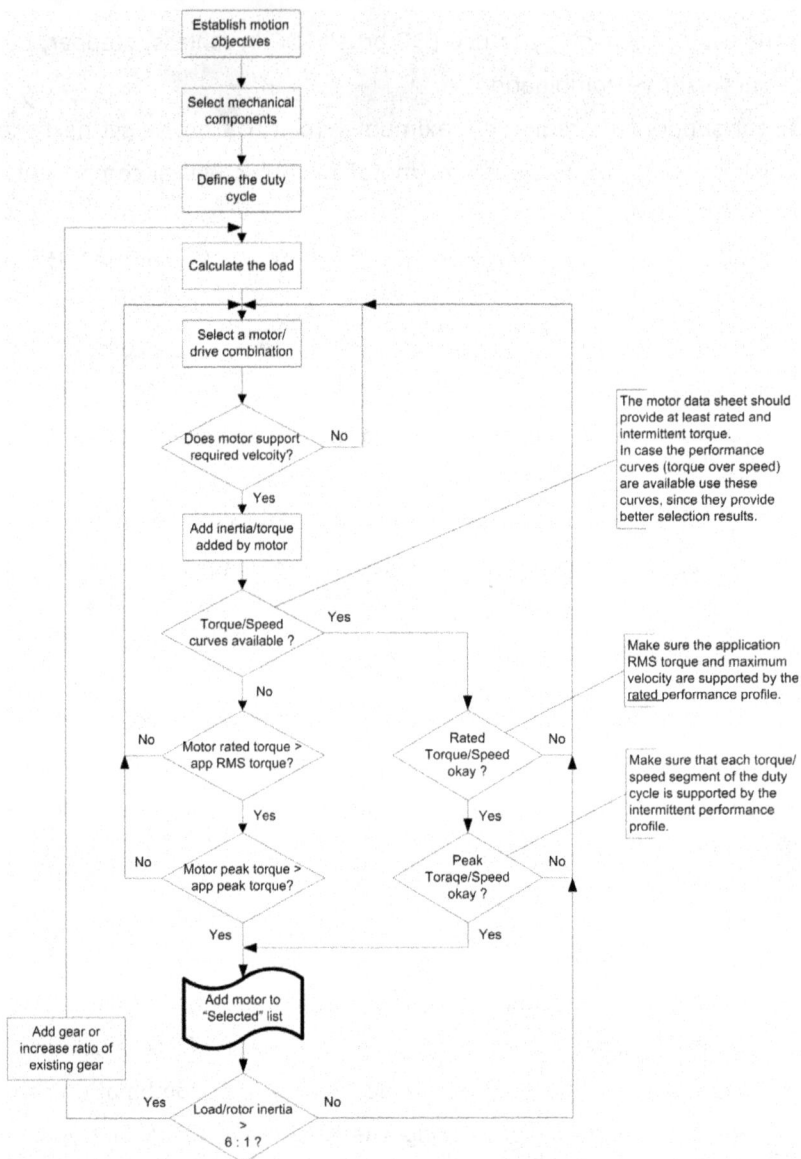

Picture 3.1: Motor Sizing and Selection Flow Chart

The following chapters will explain the steps between 'Selection of mechanical components' and 'Motor Selection' in more detail, supported by a sample application.

3.1   Selection of mechanical components

The engineer must decide which mechanical components are required for the application. For instance, a linear application may require a leadscrew or a conveyor. For speed transmission a gear or a belt drive may be used.

➢ Direct Drive?

➢ Special application or standard mechanical devices ?

➢ If linear application: Use of linear motor or leadscrew, conveyor, etc.?

➢ Reducer required – Gearbox, belt drive, etc.?

➢ Check shaft dimensions – select couplings

➢ Check mechanical components for speed and acceleration limitations

The most common mechanical components for a motion application are:

➢ **For speed transmissions:**
  - Gear
  - Belt Drive
  - Chain Drive (Chain-Sprocket)

➢ **For linear movements:**
  - Conveyor
  - Leadscrew
  - Linear Actuator[2]
  - Rack-Pinion

➢ **For other purposes:**
  - Coupling
  - Brake
  - Encoder

➢ **Other applications:**
  - Rotary Table
  - Nip Roll
  - Winders
  - Hoist

---

[2] In all consequence, a linear actuator is actually a leadscrew (screw driven actuator) or a belt-pulley (belt driven actuator).

While a detailed functional description of these components is not in the scope of this book, it is nevertheless mandatory for motor sizing and selection to document the corresponding inertia and torque calculations. For detailed information on these calculations, refer to *Chapter 4 – Load Inertia and Torque Calculations*. The knowledge of how to calculate inertia and torque of these components will also help in the calculation of more complex mechanical units.

In the following, we will examine a very simple mechanical application, i.e. a motor and a disk as shown in the picture below.

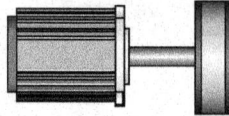

Picture 3.1.1: Disk Application

A disk actually represents the majority of motor load components, i.e. when you can calculate the inertia of a disk you can do the same for a leadscrew, conveyor, belt pulley, and many other loads. View screws, pulleys, gears, etc. as disks and hence use the same inertia equations.

Hear Ye! Hear Ye!

Sample Application:	Disk Application
1	The motion objective of this sample application shall be to accomplish a trapezoidal motion profile, i.e. to accelerate the disk to a certain speed, hold the speed for some time and then decelerate to zero speed again. The disk shall be 2" in diameter, 1.2" in thickness. The material shall be steel, which has a material density of 0.28 lbs/in^3.

3.2   Definition of a load cycle

A load cycle, i.e. the way the actual motion is applied, can have numerous shapes. There are, for instance, simple applications like blowers, conveyor drives, pumps, etc. that inflict only constant or gradually changing torque over a very long time. Sizing a motor for these applications is fairly simple and does not require further processing of the motion cycle. This book is primarily concerned with servo applications, which require abrupt and frequent torque changes during the load cycle.

The simplest forms of servo load cycles are triangular and trapezoidal motion profiles (as explained on detail in the following chapters). They define the most critical data such as maximum speed and maximum acceleration and they are sufficient enough to cover the majority of motion applications and subsequent determination of the torque requirements. Naturally, there are also very complex motion profiles and their detailed processing will result in more precise determination of the RMS torque requirement, while the peak (intermittent) torque requirement depends mainly on the maximum acceleration inside the motion cycle.

In order to process the load cycle the engineer must define the maximum velocity, maximum acceleration, duty cycle time, acceleration and deceleration ramps, dwell time, etc., specific to the application.

➢ Define critical move parameters such as velocity, acceleration rate
➢ Triangular, trapezoidal or other motion profile?
➢ If linear application: Make sure the duty cycle does not exceed the travel range of linear motion device.
➢ Jerk Limitation required?
➢ Consideration of thrust load?
➢ Does the load change during the duty cycle?
➢ Holding brake applied during zero velocity?

There are two basic (and very similar) types of a motion profile (duty/load cycle):

➢ Triangular Motion
➢ Trapezoidal Motion

3.2.1 Triangular motion profile

Picture 3.2.1.1 demonstrates the triangular motion profile:

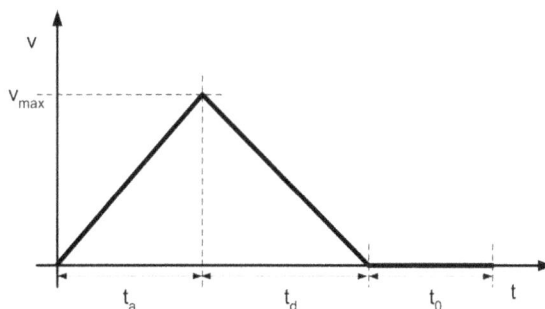

Picture 3.2.1.1: Triangular Motion Profile

15

Symbol	Description
v	Velocity
v_{max}	Maximum Velocity
t	Time
t_a	Acceleration Time
t_d	Deceleration Time
t_0	Dwell Time (Time at zero speed)

The motor accelerates to the maximum velocity and then immediately after reaching the maximum decelerates towards zero. Depending on the application, the motor may remain at standstill for some time.

note

For non-horizontal linear applications, i.e. the load is being moved in an angle up or down, it is important to consider the use of a holding brake. The motor needs to compensate for the gravitational pull on the load during zero speed cycles, which, without the use of a holding brake, will result in higher torque requirements.

3.2.2 Trapezoidal motion profile

Picture 3.2.2.1 demonstrates the trapezoidal motion profile:

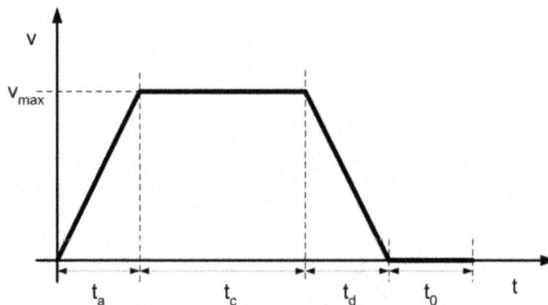

Picture 3.2.2.1: Trapezoidal Motion Profile

Symbol	Description
v	Velocity
vmax	Maximum Velocity
t	Time
tc	Constant Time
ta	Acceleration Time
td	Deceleration Time
t0	Dwell Time (Time at zero speed)

The motor accelerates to the maximum velocity, holds that velocity for some time and then decelerates towards zero. Depending on the application, the motor may remain at standstill for some time.

note For non-horizontal linear applications, i.e. moving the load in an angle up or down, it is important to consider the use of a holding brake. The motor needs to compensate for the gravitational pull on the load during zero speed cycles, which, without the use of a holding brake, will result in higher torque requirements.

3.2.3 Motion profile processing

note The following equations are universal between triangular and trapezoidal motion profiles, considering that a triangular motion profile behaves like a trapezoidal motion profile without the constant time (time at constant speed).

In order to calculate the torque requirements we need the following data from the motion profile:

➢ **RMS Torque**
- Total cycle time
- Acceleration/Deceleration time
- Constant time (time at constant speed; will be zero for triangular profile)
- Dwell time (time at zero speed)

> ➢ **Peak (Intermittent) Torque**
>> • Maximum Acceleration/Deceleration (Torque = Inertia times Acceleration)

The duty cycle parameters for the determination of the RMS torque can naturally be derived directly from the motion profile. The maximum acceleration is calculated as shown below:

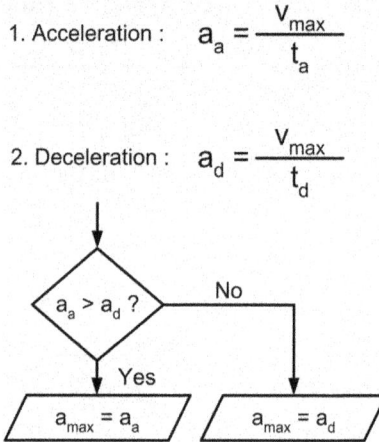

$$1.\ \text{Acceleration}: \quad a_a = \frac{V_{max}}{t_a}$$

$$2.\ \text{Deceleration}: \quad a_d = \frac{V_{max}}{t_d}$$

$$a_a > a_d\ ? \quad \xrightarrow{\text{No}} \quad a_{max} = a_d$$

$$\downarrow \text{Yes}$$

$$a_{max} = a_a$$

Picture 3.2.3.1: Determination of maximum acceleration

note In order to determine the maximum acceleration/deceleration it is necessary to use the absolute amount of the deceleration, since deceleration is basically a negative acceleration. The maximum torque will occur during the highest acceleration/deceleration.

In case a more complex motion profile is required, the engineer will need to process all time segments in order to calculate the RMS torque. To calculate the peak (intermittent) torque the engineer needs to record the acceleration/deceleration of each time segment and determine the maximum acceleration from these values as shown in the next example.

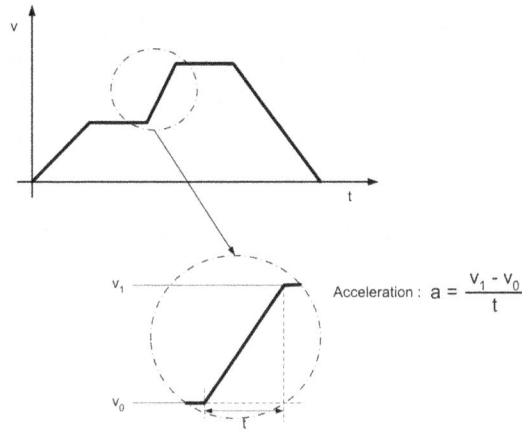

Picture 3.2.3.2: Acceleration calculation for complex motion profile

note Some applications may require different deceleration ramps, for instance, one for regular deceleration (regular stop command) and one for emergency operation (emergency stop command). In such a case, the emergency stop deceleration may determine the highest torque requirement.

The following picture demonstrates the difference between a triangular and a trapezoidal motion profile in terms of torque requirements.

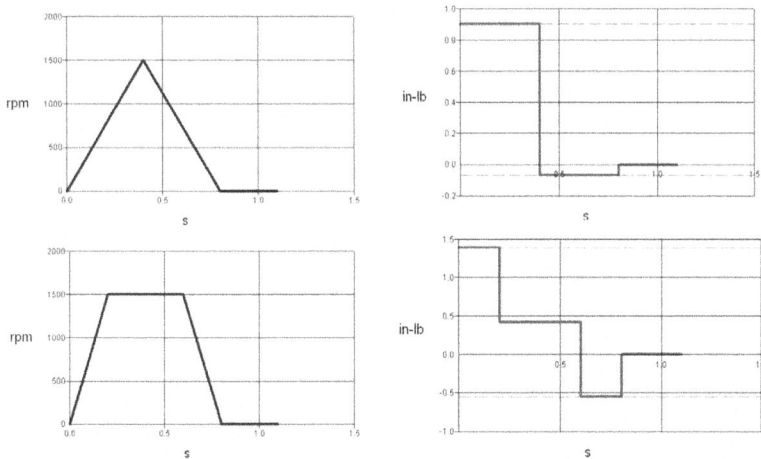

Picture 3.2.3.3: Torque during Triangular and Trapezoidal Motion Profile

Both motion profiles use the same total cycle time. The trapezoidal profile, however, requires a higher acceleration/deceleration rate, which in turn results in higher torque requirements. This circumstance can be of importance for some motion control applications.

Sample Application:	Calculation of Maximum Load Acceleration
2	The motion profile shall be of trapezoidal shape using the following parameters: $t_a = t_d = 1$ sec $t_c = 2$ sec $t_0 = 1$ sec $v_{max} = 1000$ rpm $= 16.67$ rev/sec The derived parameters are: $t_{total} = 4$ sec $a_{max} = v_{max} / t_a = 16.67 / 1 = 16.67$ rev/sec^2

3.2.4   Motion profile calculation

The following chapter provides some more insight into the calculation of the motion profile in a more generic sense. The equations as shown here are based on the use of radians for traveled distance, radians/sec for speed and radians/sec^2 for acceleration and deceleration.

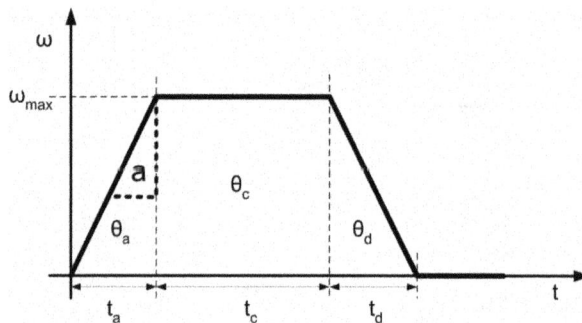

Picture 3.2.4.1: Trapezoidal Motion Profile

Symbol	Description	Angular unit	Rotational unit
ω	Velocity	rad/sec	rps or rpm
ωmax	Maximum Velocity	"	"
a	Acceleration	rad/sec2	rev/ sec2
Θ	Distance (rotation)	rad	rev
Θa	Distance (rotation) during acceleration	"	"
Θc	Distance (rotation) during constant speed	"	"
Θd	Distance (rotation) during deceleration	"	"
t	Time	sec	sec
tc	Constant Time	"	"
ta	Acceleration Time	"	"
td	Deceleration Time	"	"

The equations for trapezoidal moves are:

$$\theta_{Total} = \theta_a + \theta_c + \theta_d = \omega_{max} \times \left(\frac{t_a}{2} + t_c + \frac{t_d}{2} \right)$$

$$\omega_{max} = \frac{\theta_{Total}}{\left(\frac{t_a}{2} + t_c + \frac{t_d}{2} \right)}$$

Equation 3.2.4.1: Trapezoidal Moves

The equations for triangular moves (with t_c = 0) are:

$$\theta_{Total} = \theta_a + \theta_d = \omega_{max} \times \left(\frac{t_a}{2} + \frac{t_d}{2} \right)$$

$$\omega_{max} = \frac{\theta_{Total}}{\left(\frac{t_a}{2} + t_c + \frac{t_d}{2} \right)}$$

With $t_a = t_d$:

$$\omega_{max} = \frac{\theta_{Total}}{t_a}$$

Equation 3.2.4.2: Triangular Moves

Knowing that the area under the velocity vs. time segments represents the traveled distance and their slopes are the acceleration makes these equations easily remembered. Thusly, the equations are based on the area calculations of rectangles and triangles and their angles, respectively.

$$\alpha = \frac{\left(\omega_{max} - \omega_0\right)}{t_a} \times 2\pi$$

Equation 3.2.4.3: Acceleration calculation

ω_0 represents the initial angular or rotational velocity. With $\omega_0 = 0$ the equation changes to:

$$\alpha = \frac{\omega_{max}}{t_a} \times 2\pi$$

Equation 3.2.4.4: Acceleration calculation with $\omega_0 = 0$

Naturally, there are motion applications that require a more complex load cycle and subsequent processing of the motion profile, which in turn has major implications on the load torque. These applications include, for instance, vertical movements, jerk limitation (S-Curve), etc. which will be explained later in more detail.

Hear Ye! Hear Ye!

3.2.5   Motion profile equations

The following provides some common motion equations such a speed, acceleration, etc. The parameters used are:

Parameter	Description	English Engineering	Metric
θ	Rotation	rev	rad
ω	Velocity	rev/sec, rpm	rad/sec
t	Time	sec	sec
α	Acceleration	rev/sec^2	rad/sec^2

Unknown	Known	Equation
θ [radians]	ω_0, t, α	$\theta = (\omega_0 \times t) + \dfrac{at^2}{2}$
	$\omega_{max}, \omega_0, t$	$\theta = (\omega_{max} + \omega_0) \times \dfrac{t}{2}$
	$\omega_{max}, \omega_0, \alpha$	$\theta = \dfrac{\omega_{max}^2 - \omega_0^2}{2 \times \alpha}$
	ω_{max}, t, α	$\theta = (\omega_{max} \times t) - \dfrac{\alpha \times t^2}{2}$
ω_{max} [rad/sec]	ω_0, t, α	$\omega_{max} = \omega_0 + (\alpha \times t)$
	θ, ω_0, t	$\omega_{max} = \dfrac{2 \times \theta}{t} - \omega_0$
	θ, ω_0, α	$\omega_{max} = \sqrt{\omega_0^2 + (2 \times \alpha \times \theta)}$
	θ, t, α	$\omega_{max} = \dfrac{\theta}{t} + \dfrac{\alpha \times t}{2}$
ω_0 [rad/sec]	ω_{max}, t, α	$\omega_0 = \omega_{max} - (\alpha \times t)$
	θ, ω_{max}, t	$\omega_0 = \dfrac{2 \times \theta}{t} - \omega_{max}$
	$\theta, \omega_{max}, \alpha$	$\omega_0 = \sqrt{\omega_{max}^2 - (2 \times \alpha \times \theta)}$
	θ, t, α	$\omega_0 = \dfrac{\theta}{t} - \dfrac{\alpha \times t}{2}$
t [sec]	$\omega_{max}, \omega_0, \alpha$	$t = \dfrac{\omega_{max} - \omega_0}{\alpha}$
	$\theta, \omega_{max}, \omega_0$	$t = \dfrac{2 \times \theta}{\omega_{max} - \omega_0}$
α [rad/sec²]	$\theta, \omega_{max}, \omega_0$	$\alpha = \dfrac{\omega_{max}^2 - \omega_0^2}{2 \times \theta}$
	$\omega_{max}, \omega_0, t$	$\alpha = \dfrac{\omega_{max} - \omega_0}{t}$
	θ, ω_0, t	$\alpha = 2 \times \left(\dfrac{\theta}{t^2} - \dfrac{\omega_0}{t} \right)$
	θ, ω_{max}, t	$\alpha = 2 \times \left(\dfrac{\omega_{max}}{t} - \dfrac{\theta}{t^2} \right)$

3.2.6   Jerk Limitation

Jerk Limitation (a.k.a. S-Curve Profiling) is the time rate of change of axis acceleration/deceleration. It provides smoother motion control by reducing the jerk (rate of change) in acceleration and deceleration portions of the motion profile.

Its purpose is to eliminate mechanical jerking when speed changes are made and can be used to minimize mechanical wear and tear, optimizing travel response, reduce splashing when transporting liquids, prevent tipping when transporting boxes on a conveyor, etc.

Jerk limitation is defined as a percentage of each deceleration and acceleration segment. This percentage is then equally divided between the beginning and end of the segment. For example, for an acceleration segment of 1 second and a jerk limitation of 50%, the first 25% (from 0 to 0.25 seconds) and last 25% (from 0.75 to 1 second) of the segment include jerk limitation. A value of zero is used for linear acceleration.

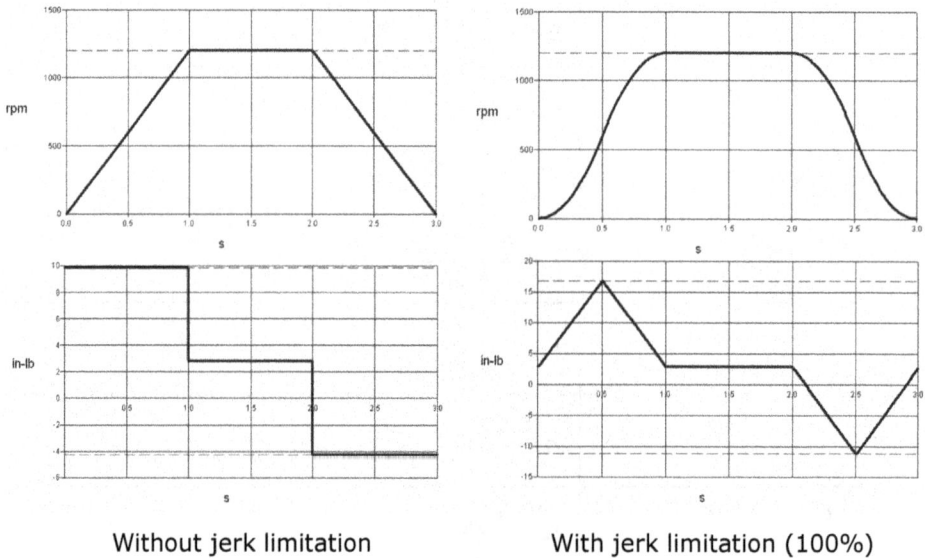

Without jerk limitation   With jerk limitation (100%)

Picture 3.2.6.1: **Speed and Torque Profile**

Without jerk limitation   With jerk limitation (100%)

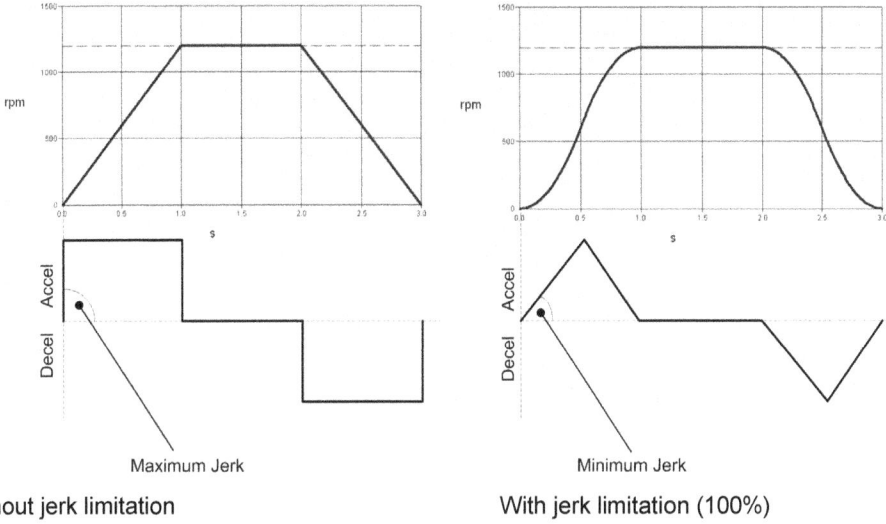

Picture 3.2.6.2: Velocity and Acceleration

Picture 3.2.6.2 demonstrates clearly the difference in acceleration between linear acceleration (no jerk limitation) and S-Curve profiling (with jerk limitation).

note Jerk Limitation increases the torque requirements and may result in a need for a larger motor, i.e. larger jerk factors require a larger peak torque from the motor. For example, a segment with 100% jerk limitation requires twice the peak (acceleration) torque as the same segment using linear acceleration.

The following series of pictures shows a comparison of different jerk limitation percentages and the resulting torque requirements:

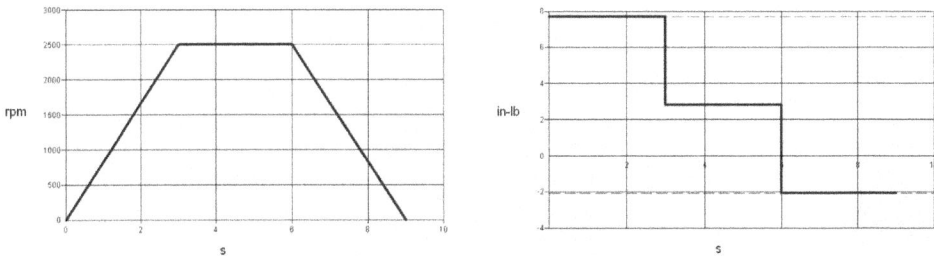

Picture 3.2.6.3: 0% Jerk Limitation

25

Note that a value of 0% results in a linear motion and torque profile. The intermittent (peak) torque requirement in this example is not quite 8 in-lbs.

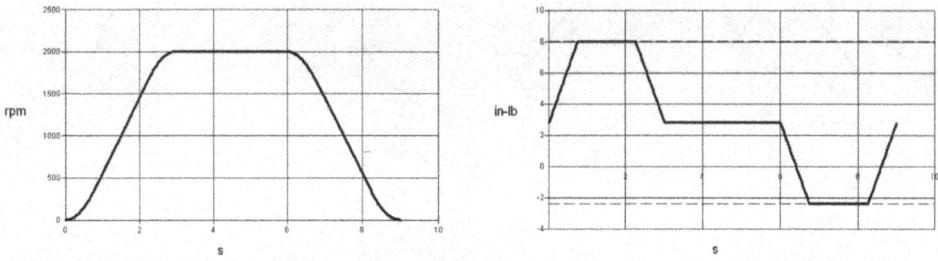

Picture 3.2.6.4: 50% Jerk Limitation

This is the same example as shown previously, but in this case with a jerk limitation of 50%. The effects of the jerk limitation are visible. Note that the torque profile shows a trapezoidal shape. Also note that the intermittent torque requirement is slightly higher than in the previous picture, i.e. a little over 8 in-lbs.

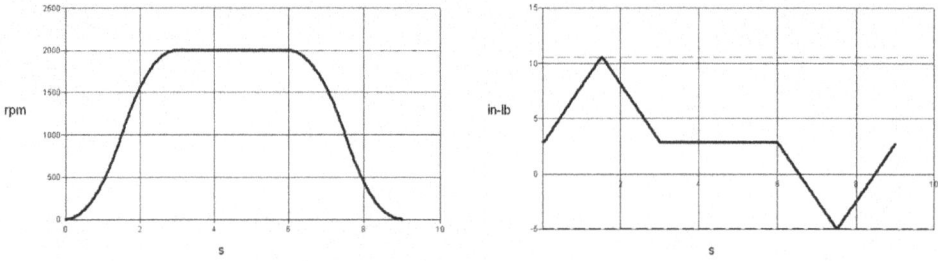

Picture 3.2.6.5: 100% Jerk Limitation

Picture 3.2.6.5 shows the effects of 100% jerk limitation (using the same sample application as the previous pictures). Note that the torque profile shows a triangular shape. Also note the significantly higher torque requirements of roughly 11 in-lbs.

note

Be aware that the previous sample application requires a constant torque of roughly 3 in-lbs. The previous statement that 100% jerk limitation requires twice the peak torque than a linear acceleration remains true.

Last, but not least, picture 3.2.6.6 demonstrates why the intermittent torque requirements are higher with jerk limitation:

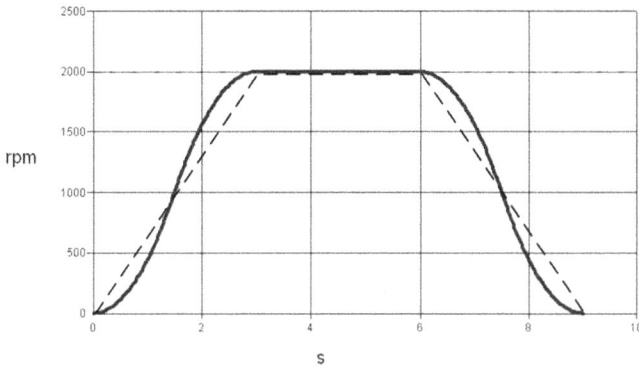

Picture 3.2.6.6: Linear and jerk limited motion profile

The jerk limited motion profile requires a higher acceleration and deceleration rate than the linear profile and higher acceleration/deceleration results directly into higher torque.

3.2.6.1   S-Curve Calculation

The following equations are based on the assumption that the area under an acceleration vs. time graph is the equivalent of the velocity, since $v = a \times t$. Let's have a look at the acceleration ramp of a sample S-Curve and its resulting acceleration graph.

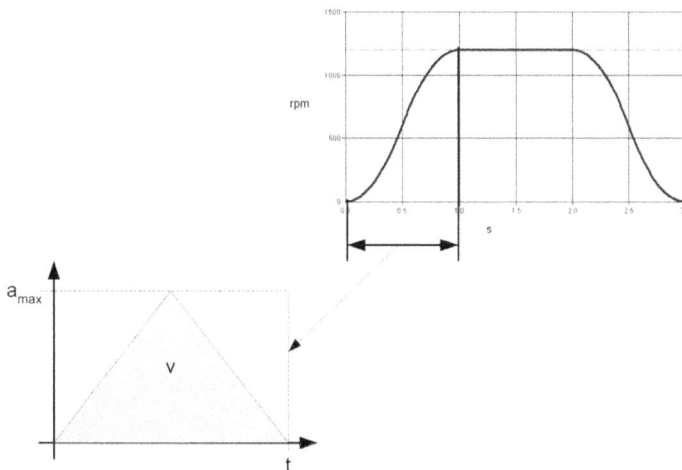

Picture 3.2.6.1.1: Acceleration ramp

The sample shows a velocity profile with 100% jerk limitation. The final velocity based on picture 3.2.6.1.1. can be calculated as follows:

$$v_f = \frac{a_{max} \times t}{2}$$

v_f	Final Velocity
a_{max}	Max. Acceleration
t	Total Acceleration Time

Equation 3.2.6.1.1: S-Curve Final Velocity

However, in order to cover all possible jerk limitation percentages we need to look at a different acceleration curve. The following picture demonstrates a generic acceleration ramp.

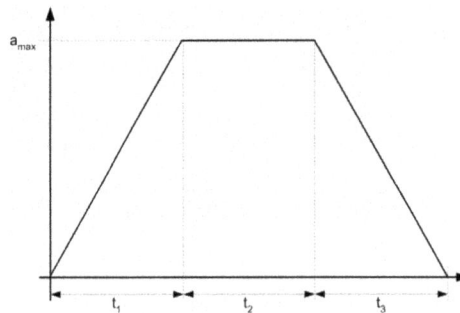

t_1	Time of Increasing Acceleration
t_2	Time of Constant Acceleration
t_3	Time of Decreasing Acceleration
$t_1 = t_3 = 0$	Jerk Limitation = 0%
$t_2 = 0$	Jerk Limitation = 100%

Picture 3.2.6.1.2: Generic S-Curve Acceleration Ramp

Picture 3.2.6.1.2 shows an acceleration ramp that represents roughly a 60% jerk limitation. As was explained previously, the percentage is equally divided between the beginning and end of the segment. For example, for an acceleration segment of 1 second and a jerk limitation of 60%, the first 30% (from 0 to 0.3 seconds) and last 30% (from 0.7 to 1 second) of the segment include jerk limitation. A value of zero is used for linear acceleration.

Based on this definition of the jerk percentage it can be safely assumed that t_1 and t_3 have equal values. The final velocity for the generic S-Curve acceleration ramp can be calculated as follows:

$$v_f = \frac{a_{max} \times t_1}{2} + a_{max} \times t_2 + \frac{a_{max} \times t_3}{2}$$

With $t_1 = t_3$:

$$v_f = a_{max} \times t_1 + a_{max} \times t_2$$

$$v_f = a_{max} \times (t_1 + t_2)$$

Equation 3.2.6.1.2: S-Curve Final Velocity (Generic)

The next assumption is that the motion control engineer tends to define the maximum velocity for a given application and that, in order to create a velocity profile; the equations should provide the information of the velocity at any given time.

With the final velocity v_f known and also assuming that the acceleration/deceleration ramp times are known we can now calculate the maximum acceleration.

$$a_{max} = \frac{v_f}{t_1 + t_2}$$

Equation 3.2.6.1.3: S-Curve Maximum Acceleration

In the following, we need to look at three different sections of the acceleration ramp (See also picture 3.2.6.1.2):

➢ Increasing Acceleration
➢ Constant Acceleration
➢ Decreasing Acceleration (Deceleration)

1. Increasing Acceleration: $0 \le t \le t_1$

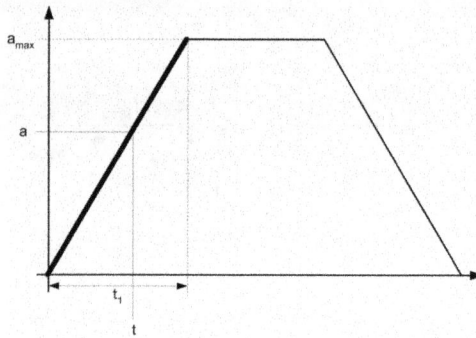

Picture 3.2.6.1.3: Calculation of Velocity During Increasing Acceleration

1. Acceleration at any given time t:

$$a = a_{max} \times \frac{t}{t_1}$$

2. Velocity v_t at any given time t:

$$v_t = \frac{a \times t}{2}$$

$$v_t = \frac{a_{max} \times \dfrac{t}{t_1} \times t}{2}$$

$$v_t = \frac{a_{max} \times t^2}{2 \times t_1}$$

2. Constant Acceleration: $t_1 \le t \le (t_1 + t_2)$

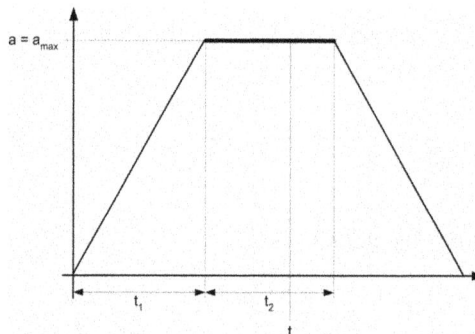

Picture 3.2.6.1.4: Calculation of Velocity During Constant Acceleration

1. Acceleration at any given time t: $a = a_{max}$

2. Adjust time: $t = t - t_1$

3. Velocity v_t at any given time t:

$$v_t = \frac{a_{max} \times t_1}{2} + a_{max} \times t$$

$$v_t = a_{max} \times \left(\frac{t_1}{2} + t \right)$$

3. Decreasing Acceleration (Deceleration): $(t_1 + t_2) \le t \le (t_1 + t_2 + t_3)$

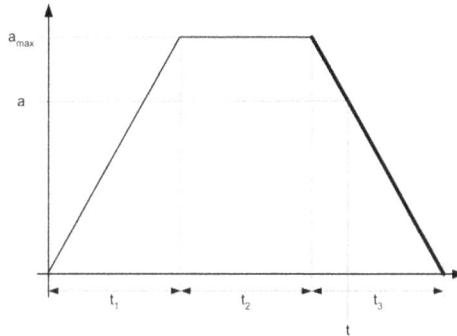

Picture 3.2.6.1.5: Calculation of Velocity During Decreasing Acceleration

1. Acceleration at any given time t:

$$a = a_{max} \times \left(1 - \frac{t}{t_3} \right)$$

2. Adjust time:

$$t = t - t_1 - t_2$$

3. Velocity v_t at any given time t:

$$v_t = \frac{a_{max} \times t_1}{2} + a_{max} \times t2 + \frac{a \times t}{2}$$

$$v_t = \frac{a_{max} \times t_1}{2} + a_{max} \times t2 + \frac{a_{max} \times \left(1 - \frac{t}{t_3} \right) \times t}{2}$$

$$v_t = a_{max} \times \left(\frac{t_1}{2} + t_2 + \frac{t - \frac{t^2}{t_3}}{2} \right)$$

Last, but not least, let's accomplish some calculations of a sample application profile in order to prove the validity of the previously described equations.

The sample profile is based on the following parameters:

Max Velocity	1000 rpm
Acceleration Ramp Time	100 msec
Jerk Limitation	50%
t1	25 msec
t2	50 msec
t3	25 msec
Max. Acceleration	13.33 rev/msec^2
	(Calculated; see equation 3.2.6.1.3)

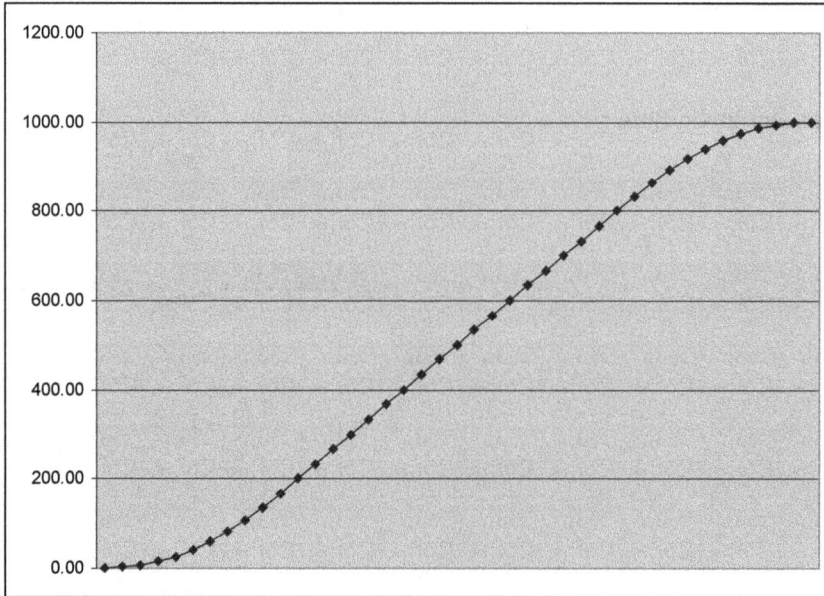

Picture 3.2.6.1.6: Proof of S-Curve Equations through MS-Excel

The calculated velocity points are as follows (based on 2.5 msec increments):

	t	v_t
1. Increasing Acceleration	0	0.00
	2.5	1.67
	5	6.67
	7.5	15.00
	10	26.67
	12.5	41.67
	15	60.00
	17.5	81.67
	20	106.67
	22.5	135.00
	25	166.67
2. Constant Acceleration	27.5	200.00
	30	233.33
	32.5	266.67
	35	300.00
	37.5	333.33
	40	366.67
	42.4	400.00
	45	433.33
	47.5	466.67
	50	500.00
	52.5	533.33
	55	566.67
	57.5	600.00
	60	633.33
	62.5	666.67
	65	700.00
	67.5	733.33
	70	766.67
	72.5	800.00
	75	833.33
3. Decreasing Acceleration:	77.5	865.00
	80	893.33
	82.5	918.33
	85	940.00
	87.5	958.33
	90	973.33
	92.5	985.00
	95	993.33
	97.5	998.33
	100	1000.00

3.3   Load calculation

The load of a motor is determined by its own rotor inertia, the total inertia reflected from the mechanical set up, the constant torque from the mechanical set up, and the maximum speed and maximum acceleration of the application.

As demonstrated in the following picture any calculation of the load parameters starts with the last mechanical component in the set up.

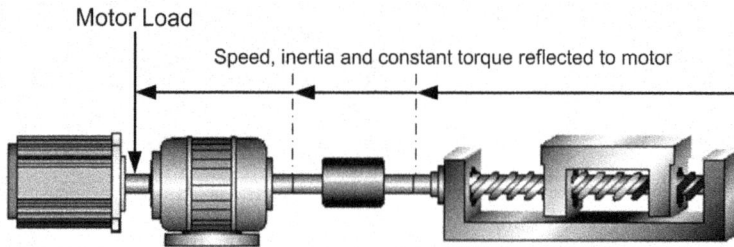

Picture 3.3.1: Motor load

The speed, acceleration, inertia and constant torque is "reflected" from mechanism to mechanism until it reaches the motor. Each component will add its own inertia and constant torque. Speed transmission components, like, for instance, a gear, will also transform the inertia, speed and acceleration from the previous component according to the transmission ratio.

The total (system) inertia and the maximum acceleration will determine the acceleration torque. The total torque, i.e. the required peak (intermittent) torque is the sum of acceleration torque and constant torque. The required RMS torque result from the entire motion cycle and the acceleration and constant torque of each motion segment as will be explained in more detail later in this chapter.

The following picture explains the process in more detail using the same sample application.

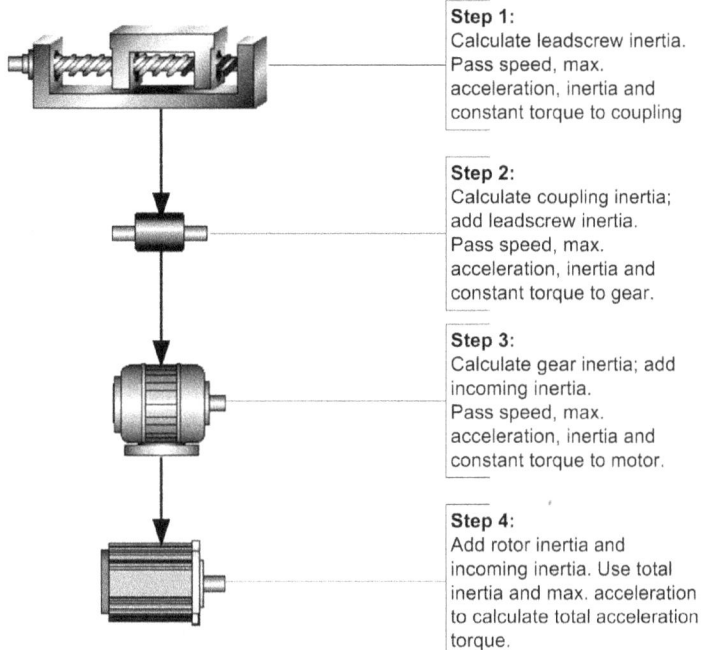

Step 1:
Calculate leadscrew inertia. Pass speed, max. acceleration, inertia and constant torque to coupling

Step 2:
Calculate coupling inertia; add leadscrew inertia. Pass speed, max. acceleration, inertia and constant torque to gear.

Step 3:
Calculate gear inertia; add incoming inertia. Pass speed, max. acceleration, inertia and constant torque to motor.

Step 4:
Add rotor inertia and incoming inertia. Use total inertia and max. acceleration to calculate total acceleration torque.

Picture 3.3.2: Motor load transmission

The basic required data to select a motor are:

➢ Load maximum speed

➢ Load maximum (intermittent) torque

➢ Load RMS (rated) torque

➢ Load inertia

The following shows an excerpt from a report created with VisualSizer-Professional. The application comprises of a motor, a servo reducer and a leadscrew. The leadscrew moves a weight of 3 lbs.

Selected Motor:

Manufacturer		AutomationDirect.Com
Product Family		SVL - Low Inertia Motors
Product Key		SVL-204
Drive Module		SVA-2040
Rated Speed	3000	rpm
Rated Torque	11.236	in-lb
Max. (Peak) Torque	33.798	in-lb
Rotor Inertia	0.00030082	in-lb-s^2
Kt	0	in-lb/A

Motor Load Data:

Max. Velocity	3000	rpm
Constant Torque	2.8353	in-lb
RMS Torque	5.2295	in-lb
Peak Torque	10.525	In-lb
Load Inertia	0.00247237	in-lb-s²
Ratio Load/Rotor Inertia	8.2188	: 1

System Data (incl. Rotor Inertia):

RMS Torque	5.7357	in-lb
Peak Torque	11.46	in-lb
Total Inertia	0.00277319	in-lb-s²

Mechanism No. 1: Right-Angle Servo Reducer

Manufacturer		SHIMPO DRIVES, INC.
Product Key		NEVKFE15P20019011LR
Reducer Ratio	15	: 1
Moment of Inertia	0.002472	in-lb-s²
Internal Losses	2.777	in-lb
Reducer Efficiency	92	%
Weight	27.6	lb
Nominal Output Torque	956	in-lb
Max. Output Torque	1744	in-lb
Max. Input Speed	5000	rpm

Calculated Input Shaft Load Data:

Max. Input Velocity	3000	rpm
Total Inertia	0.00247237	in-lb-s²
Mechanism Inertia	0.002472	in-lb-s²
Constant Torque	2.8353	in-lb
Peak Torque	10.525	in-lb

Calculated Output Data:

RMS Torque at Output Shaft	0.685827	in-lb
Peak Torque at Output Shaft	0.820157	in-lb

Mechanism No. 2: Leadscrew

Manufacturer		Blue Sky Lead Screws
Product Key		BSLS 001
Screw Diameter	0.35	in
Screw Length	25	in
Material Density	0.28	lb/in³
Pitch / Lead		25
rev/in		
Mechanism Efficiency	96.8	%
Friction Coefficient	0.11	

Preload Torque	0.8	in-lb
Thrust Forces	0	lb
Table Weight	2.7	lb
Angle	0	°
Coupling Inertia	0.00005	in-lb-s²
Max. Velocity	0	rpm

Calculated Input Shaft Load Data:

Max. Input Velocity	200	rpm
Total Inertia	0.00007733	in-lb-s²
Mechanism Inertia	0.000027	in-lb-s²
Constant Torque	0.804124	in-lb
Peak Torque	0.820157	in-lb

Load Parameters:

Type		Weight Only
Weight	3	lb

The calculation of the motor load is accomplished as follows:

1. Add load weight to the leadscrew table weight.

2. Calculate the screw inertia and add the table & load inertia reflected to the input shaft.

3. Apply the resulting inertia as load inertia to the servo reducer.

4. Add the servo reducer inertia plus the load inertia reflected to the input shaft and apply it to the motor.

In order to apply speed and acceleration to the motor shaft take into account the transmission ratio (or lead/pitch) of each device.

As was mentioned earlier, there are simple applications like blowers, conveyor drives, pumps, etc. that inflict only constant or gradually changing torque over a very long time. Sizing a motor for these applications is fairly simple and does not require any complex processing of the motion cycle. To size a motor for this kind of application it is only necessary to match the load torque, which is usually accomplished by matching the load with the motor's horsepower using the following equation:

$$P_{[HP]} = \frac{v_{[rpm]} \times T_{[ft-lb]}}{5250}$$

1 HP = 746 Watts = 550 ft-lb/sec

3.3.1 Load maximum speed

The motion profile directly acquires the maximum speed as demonstrated in the example below, where v_2 represents the maximum speed.

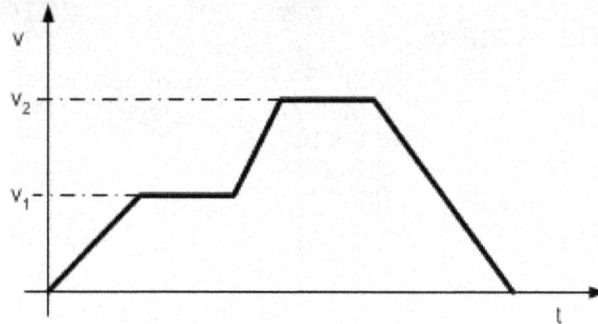

Picture 3.3.1.2: Determination of maximum speed

In regards to the sample application, we apply the maximum speed as shown below.

Sample Application:	Determination of Maximum Load Speed
3	As defined in the motion profile the maximum required speed is: v_{max} = 1000 rpm

3.3.2 Load inertia and maximum torque

The calculation of the maximum (intermittent) torque is fairly easy, since it depends mainly on the maximum acceleration. However, the maximum torque contains of two components:

1. Constant torque as inflicted by the mechanical setup. This is the torque due to all other non-inertial forces such as gravity, friction, preloads and other push-pull forces.
2. Acceleration torque as inflicted by the inertia[3] from the mechanical setup plus the maximum acceleration as required by the load cycle.

---

[3] For a detailed documentation on inertia and torque calculations please refer to *Chapter 4 – Inertia and Torque Calculation.*

note

Be aware that, during the motor selection process, the motor's rotor inertia must be added to the load inertia.

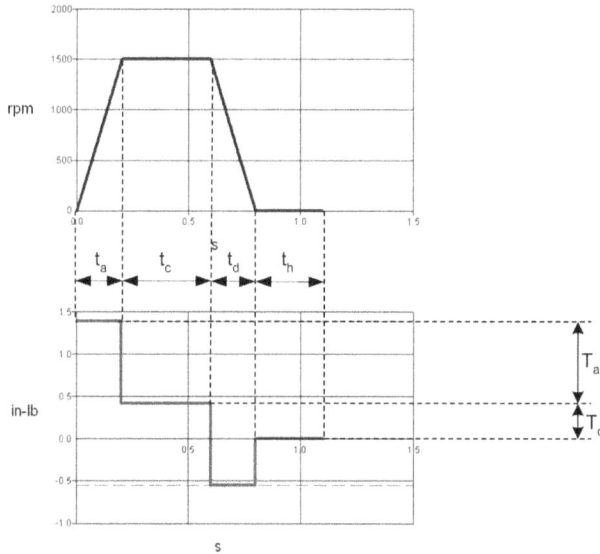

Picture 3.3.2.1: Example of acceleration and constant torque

The equations used to calculate the maximum torque are:

$$J_{system} = J_{app} + J_{rotor}$$

$$T_a = J_{system} \times a$$

$$T_{total} = T_a + T_c$$

Equation 3.3.2.1: Determination of maximum torque

Symbol	Description
J_{system}	Total system inertia
J_{app}	Application inertia (inflicted by mechanical setup)
J_{rotor}	Rotor inertia
T_a	Acceleration torque
a	Acceleration
T_{total}	Total (system) torque
T_c	Constant torque (inflicted by mechanical setup)

For the sample application, we apply the following:

Sample Application:	Calculation of Intermittent/Peak Torque
4	Max. acceleration a_{max} = 16.67 rev/sec² as derived from the motion profile. The inertia of the disk is determined as: Disk Weight: $W = \pi \times \left(\dfrac{D}{2}\right)^2 \times L \times \rho$ Disk Inertia: $J_{disk} = \dfrac{W \times r^2}{2 \times g} = \dfrac{\pi \times L \times \rho \times r^4}{2 \times g}$ Using the disk dimensions as defined earlier the disk inertia will be: $J_{disk} = 0.00137 in - lb - \sec^2$ In the following, we assume a rotor inertia of 0.0011 in-lb-sec². $J_{system} = 0.00137 + 0.0011 = 0.00247 in - lb - \sec^2$ Since a disk does not inflict any friction, we assume the constant torque to be zero. As a result the acceleration torque represents the total torque: $T_a = T_{total} = J_{system} \times \alpha \times 2 \times \pi = 0.259 in - lb$

Symbol	Description
W	Weight
Π	"PI" = 3.14159
D	Disk diameter
L	Disk thickness (length)
ρ	Material density
J_{disk}	Disk inertia
r	Disk radius
g	Gravity constant (386 in/sec^2)

The calculation of the disk's inertia as demonstrated with the sample application provides a first glimpse into the inertia calculation of, in this case, a basic component. The calculation of, for instance, the inertia of a leadscrew, a rotary table, a pulley, etc. is based on the same math as shown here.

3.3.3 Load RMS torque

While the calculation of the maximum (intermittent) torque is fairly easy, the calculation of the RMS load torque is a bit more complex. The RMS torque ("Root Mean Squared") represents the average torque over the entire duty cycle. To calculate the RMS torque the following parameters are required:

- ➢ Torque during acceleration
- ➢ Torque during constant speed
- ➢ Torque during deceleration
- ➢ Acceleration time
- ➢ Time of constant speed
- ➢ Deceleration time
- ➢ Holding time

The following picture demonstrates the determination of these parameters.

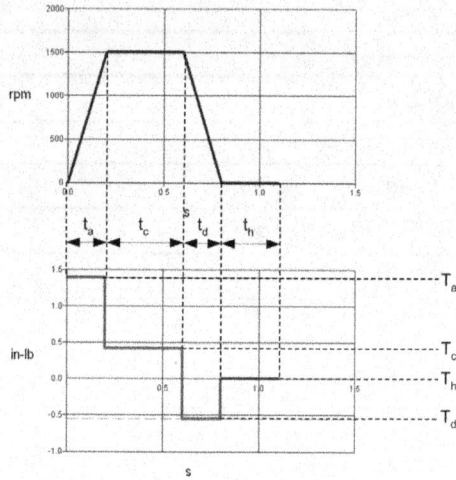

Picture 3.3.3.1: Demonstration of torque and time segments

Symbol	Description
Ta	Total torque during acceleration (incl. constant torque)
Tc	Constant torque due to friction
Td	Total torque during deceleration (incl. constant torque)
ta	Acceleration time
tc	Time of constant speed
td	Deceleration time
th	Holding time

note

T_a, T_c, T_d and T_h as shown above refer to the total torque, i.e. including constant torque, during acceleration, constant time and deceleration, while in the following equation the constant torque is handled separately.

The equation to calculate the RMS torque is as follows:

$$T_{RMS} = \sqrt{\frac{(T_a + T_c)^2 \times t_a + T_c^2 \times t_c + (T_d + T_c)^2 \times t_d + T_h^2 \times t_h}{t_a + t_c + t_d + t_h}}$$

Equation 3.3.3.1: RMS torque calculation

Symbol	Description
Ta	Acceleration torque
Tc	Constant torque
Td	Deceleration torque
Th	Holding torque
ta	Acceleration time
tc	Time of constant speed
td	Deceleration time
th	Holding time

note

T_a, T_c, T_d and T_h as shown in the equation represent absolute values. Deceleration torque is usually negative. The motor, however, has to provide at least that amount of torque to drive the mechanical setup during deceleration.

The equation can be simplified for a triangular motion profile (not considering the constant time t_c) as follows:

$$T_{RMS} = \sqrt{\frac{\left(T_a + T_c\right)^2 \times t_a + \left(T_d + T_c\right)^2 \times t_d + T_h^2 \times t_h}{t_a + t_d + t_h}}$$

Equation 3.3.3.2: RMS torque calculation for triangular motion profiles

The calculation of the acceleration and deceleration torque is achieved the same way as during the peak torque calculation, however, in this case the calculation has to be accomplished for each individual motion segment.

$$T_a = J_{system} \times \alpha$$

$$T_d = J_{system} \times d$$

Equation 3.3.3.3: Determination of accel. / decel. Torque

Symbol	Description
J_{system}	Total system inertia[4]
Ta	Acceleration torque
a	Acceleration
Td	Deceleration torque
d	Deceleration

note

Acceleration means in this case angular acceleration. Angular acceleration is measured in radians per $\sec^2$.

The constant torque, i.e. the torque due to all other non-inertial forces such as gravity, friction, preloads and other push-pull forces, must be derived from component data sheets (e.g. a leadscrew's preload torque) or application specifics (refer to *Chapter 4 – Load Inertia and Torque Calculations*).

Things become a bit more tedious in cases where a motion profile more complex than triangular or trapezoidal is required as demonstrated in the following picture.

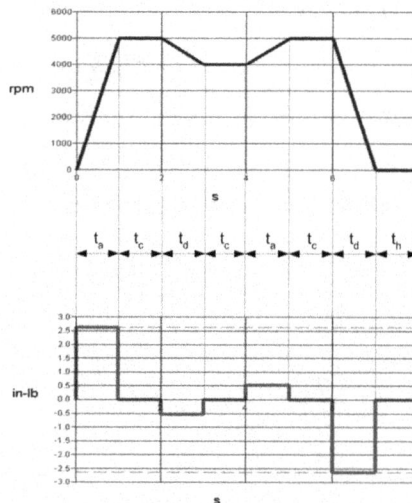

Picture 3.3.3.2: Demonstration of torque and time segments

---

[4] For the calculation of the system inertia refer to *Chapter 4 – Load Inertia and Torque Calculations*.

Sample Application:	Calculation of RMS Torque
5	The equation to calculate the total RMS torque for a trapezoidal motion profile, including load and motor inertia, is: $$T_{RMS} = \sqrt{\frac{\left(T_a+T_c\right)^2 \times t_a + T_c^2 \times t_c + \left(T_d+T_c\right)^2 \times t_d + T_h^2 \times t_h}{t_a+t_c+t_d+t_h}}$$ By applying our current sample application data, i.e. assuming the constant torque to be zero and acceleration/deceleration times are 1 sec, the equation can be simplified to the following: $$T_{RMS} = \sqrt{\frac{T_a^2+T_d^2}{t_a+t_c+t_d+t_h}} = \sqrt{\frac{\left(2\times 0.259\right)^2}{5}} = 0.164 in-lb$$

Symbol	Description
Ta	Acceleration torque
Tc	Constant torque
Td	Deceleration torque
Th	Holding torque
ta	Acceleration time
tc	Time of constant speed
td	Deceleration time
th	Holding time

3.4 Motor Selection

The motor selection process, as demonstrated in picture 3.4.1, can be initiated as soon as all load parameters have been established.

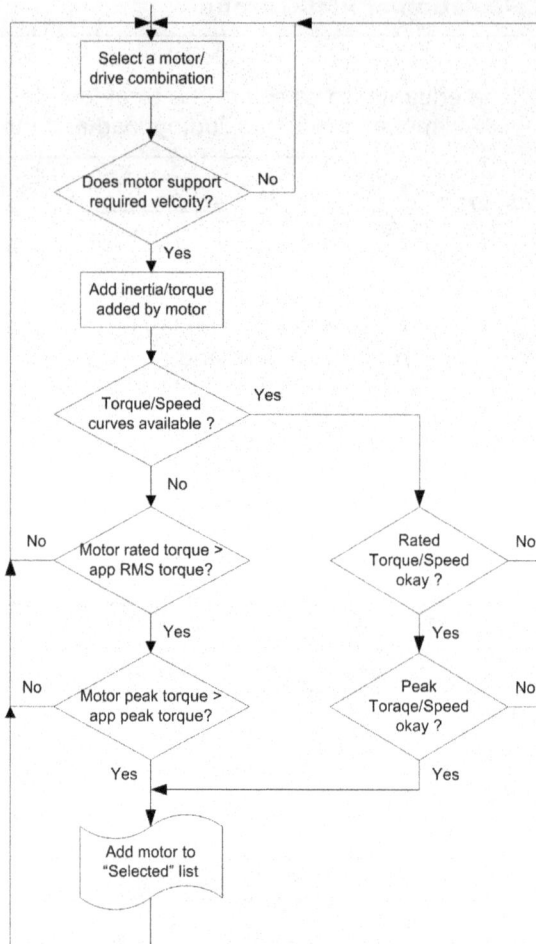

Picture 3.4.1: Motor selection process

The motor selection process as described here also explains the popularity of motor sizing programs. The process of recalculating the torque requirements for each individual motor/drive combination can be extremely time- consuming considering the vast amount of motors available in the industry. The goal of motor sizing is to find the optimum motor for the application and the only way to accomplish that is with sufficient choices available, i.e. with a great number of applicable motors.

3.4.1   Matching Motor Technologies to Applications

Prior to the actual selection process based on inertia, speed and torque comparisons, it makes sense to decide which motor technology might be best suited for the application. This method will limit the number of motors that need to be investigated and thus will reduce the time it takes to find the optimum motor.

Just as a reminder: This book is not intended to explain the basics of electrical motors. There are numerous works available on this topic. This book merely addresses the topic of motor sizing and selection. It is nevertheless necessary to address in brief the characteristics of motor technologies as they do have some bearing on the motor selection process.

The motor technologies commonly used for servo applications are:
- Stepper
- DC Brush
- DC Brushless (Permanent Magnet)
- Synchronous AC
- Asynchronous AC (AC Induction)

The DC Brushless and Synchronous AC technologies are actually identical; some manufacturers prefer to call it DC Brushless, others rather use Synchronous AC.

While the first four listed technologies are commonly used for position control, the Asynchronous AC technology is very popular for constant speed applications, but is also suitable for speed and torque control applications.

Table 3.4.1.1 provides an overview of the motor technologies and their basic operating characteristics.

Characteristic	Stepper Motor	DC Brush	DC Brushless	Asynchronous AC	Comment
Low Cost	Y	Y	Y	N	The lowest costs are usually with stepper motors or DC brush motors; the costs of brushless DC motors are slightly higher.
Smooth operation (minimal noise, vibration)	N	Y	Y	N	High-performance commutation techniques, such as sinusoidal commutation, will contribute to make brushless DC motor operation smoother.
High Speed	N	Y	Y	N	Stepper motors usually do not go over 3000 rpm.
High Power	N	N	Y	Y	Stepper motors and DC brush motors don't come usually in ranges above 1 kilo watts.
High Torque to Size Ratio	Y	N	Y	Y	Brushless DC motors provide a better spectrum of torque over speed, while stepper motor performance drops significantly at higher speeds.
Ease of Use	Y	N	N	N	No feedback and no servo tuning required.
Simplest Control Circuitry	N	Y	N	N	All motor technologies, other than DC Brush, require more than one amplifier circuit per motor.

Table 3.4.1.1: Motor Technology Characteristics

3.4.1.1   Stepper Motors

Stepper motors are basically DC brushless motors. They are self-positioning and they do not require an encoder for position feedback. However, some applications may use an encoder for the sole purpose of detecting "stall" during motion.

Stepper motors produce a high torque for a given size and weight. However, the available torque from stepper motors drops dramatically with higher speed and the complex performance curves (torque over speed) complicate the selection for a specific application. Their maximum speed is in the neighborhood of 5000 rpm at very low torque.

The power range of stepper motors is up to several hundred watts, but hardly above that level. The major drawback of stepper motors is the noise and vibration they produce. Especially vibration can affect the lifetime of a mechanical system significantly. There are, however, measures to reduce vibration such as micro stepping drive techniques or mechanical dampers, but they usually do not eliminate the problem.

3.4.1.2 DC Brush Motors

DC Brush motors are suitable for a wide variety of applications, especially for positioning, but also for speed and torque control. They do require an encoder for positioning applications. DC Brush motors are available in a great variety of sizes, up to several kilo watts. The speed range can be as high as 10,000 rpm and even higher. They run smoothly and relatively quiet.

The major disadvantage of DC Brush motors are the brushes, which do wear out over time and need to be replaced. They can also be responsible for electrical arcing. Another drawback is that DC Brush motors provide a relatively low torque in comparison to their size and weight.

3.4.1.3 DC Brushless Motors

Just like the DC Brush motor, the DC Brushless motor requires an encoder for position feedback. It is, however, the most widely used motor technology for servo applications. Brushless DC motors run relatively smooth and quiet; they do not require mechanical brushes for commutation.

Due to excellent thermodynamic features, it is able to generate high torque for a given size. DC Brushless motors are available in a wide power range and they can operate at very high speeds up to 30,000 rpm and even more. The drawback of a DC Brushless motor can be its high price due to the use of rare-earth magnetic materials to generate torque. They also require fairly complex and thus more expensive amplifiers.

3.4.1.4 AC Induction Motors

Traditionally, AC Induction motors have been used mostly for constant speed applications. Their design is simple (no magnets in either rotor or stator) and thus they work reliably over a long time; they are low-cost when they are used for constant speed applications.

Due to the emergence of more sophisticated electronic controls, which bring in a certain complexity and add to the total system costs, AC Induction motors can be used for speed and torque control applications. Although technically feasible with these electronic control techniques, AC Induction motors are rarely used for positioning applications.

3.4.2   Selection Criteria

The basic selection criteria are:

> ➤ The motor's rated speed must be equal to or exceed the application's maximum speed
> ➤ The motor's intermittent torque must be equal or exceed the load's maximum (intermittent) torque[5]
> ➤ The motor's rated torque must be equal to or exceed the load's RMS torque
> ➤ The ratio of load inertia to rotor inertia should be equal to or less than 6:1[6]

The first and most straight-forward task in the motor selection process is the comparison of the motor's rated speed with the maximum speed as required by the application.

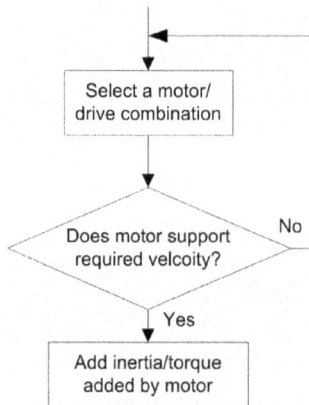

Picture 3.4.2.1: Speed comparison

---

[5] Refer to motor manufacturer data for information for what period of time the intermittent torque can be maintained by the motor.
[6] Some manufacturers state a maximum ratio of 10:1. Refer to manufacturer data for more detailed information.

However, the motor's rated speed as well as rated torque are data that can be more or less interpreted from the motor's performance profile (torque vs. speed) and they are usually determined by the manufacturer (see also chapter *3.4.1.2 Interpretation of Torque/Speed Curves*)

3.4.2.1   Inertia Matching

The generally used rule of thumb is that the armature/rotor moment of inertia should match the load moment of inertia, i.e. a ratio of 1:1 between load and motor inertia would be the ideal scenario.

The reasons to match the moment of inertia are:
1. The motor will not be able to control the speed and position of the load accurately if the load is too heavy. This will result in system instabilities such as vibrations and can lead to mechanical damages. Quick changes in speed or position become very difficult.
2. If the load inertia is too light, i.e. the motor is simply oversized for the application, most power will be used to accelerate or decelerate the rotor, rather than the load. From a power consumption standpoint, this is a waste of power and it could also lead to the overheating of the motor.

In terms of efficiency, a 1:1 ratio between load and motor inertia provides the optimum power transfer. However, a 1:1 ratio is rarely useful in an actual application, since it will require a bigger motor. From a cost savings point, especially energy consumption, this is rarely acceptable.

Hear Ye! Hear Ye!

As it turns out the inertia ratio is not that critical at all. However, one question that is being intensely discussed among experts is how high the inertia mismatch can be stretched.
A widely used rule among manufacturers is:

If the ratio of load over rotor inertia exceeds a certain range, (for servo motors 6:1) consider the use of a gearbox, increase the transmission ratio of the existing gearbox, increase the ball screw pitch, etc. This will reduce the inertia reflected to the motor. Servo motors should not be operated over a ratio of 10:1.

note

During the design process, make sure you select the gear ratio, screw pitch, etc. carefully, since this will reduce the inertia reflected to the motor and thus require smaller motors. The reflected inertia is the load inertia divided by the ratio squared.

The point to limit the range of inertia mismatch is to simplify the tuning of the system. Quick speed changes and positioning become very difficult if the mismatch ratio is too high. The load can actually drive the motor during deceleration and as a result may cause overshooting and unreasonable long settling times.

Bosch Rexroth, for instance, recommends the 'good standards' for inertia mismatch as follows:

- ➢ < 2:1 for quick positioning
- ➢ < 5:1 for moderate positioning
- ➢ < 10:1 for quick velocity changes

Much higher mismatch ratios than 10:1 can be accomplished with proper tuning of the system, however, the actual tuning process will take an unusually long time.

Hear Ye! Hear Ye!

Use the inertia ratio as an indicator of possible stability problems. However, it may not be the cause of the problem. Make sure the inertia ratio lies within a reasonable range to prevent any problems, but also check the stiffness and linearity of the system. Backlash, for instance, may cause similar problems as observed with an excessive inertia mismatch ratio.

3.4.2.2 Interpretation of Torque/Speed Curves

The performance of both motor technologies, servo brush/brushless on one side and stepper motors on the other, are most effectively demonstrated by graphs that illustrate the relationship between torque and speed, i.e. what torque is available at what speed. While stepper systems are also servo systems, there are significant differences between the performance curves of a servo and stepper motor. A profound knowledge of the differences and the proper interpretation of the performance curves is crucial for the motor selection process. It cannot be stressed enough that comparing motor data with load torque and speed requirements should be accomplished only on the basis of realistic torque vs. speed information (if such data is provided), not just rated torque and speed.

3.4.2.3 Servo Motor Performance Curves

When comparing servo motors of various manufacturers it is always advantageous to use their performance curves, i.e. torque over speed or vice versa, which in turn requires a clear understanding of these curves. For example, is the maximum torque available at zero or maximum speed? Does the curve refer to the motor alone or a motor/drive combination? What ambient temperature is assumed? There can be quite a difference in performance between 25° C and 40° C. Another factor is the voltage, which significantly affects the top speed.

Motor data derived from the performance curves and listed on the data sheet (such as rated torque, rated speed) depend very much on the manufacturer's interpretation. It can be conservative and thus leaving potential to use the motor beyond the rated parameters under certain circumstances. For marketing reasons, the interpretation may be more aggressive towards higher speed or higher torque.

The following picture represents typical torque vs. speed curves for both, brush and brushless servo systems.

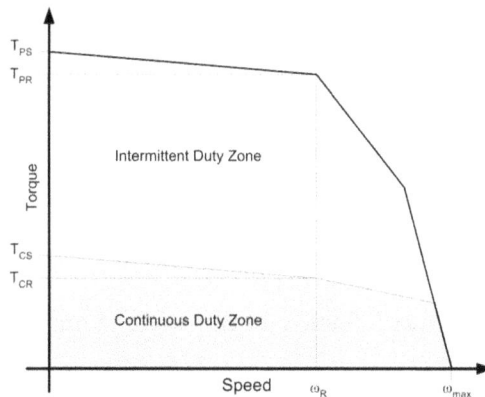

Picture 3.4.2.3.1: Typical Torque vs. Speed for Servo Motors

T_{PS} Stall peak torque
T_{PR} Rated peak torque
T_{CS} Stall continuous torque
T_{CR} Rated continuous torque
ω_R Rated speed
ω_{max} Maximum speed

53

Servos have typically two zones:

1. The continuous duty zone, in which continuous operation is possible without overheating the motor.

2. The intermittent duty zone where operation, especially acceleration, is only possible on an intermittent basis. The time that the motor or motor/drive combination can maintain the intermittent torque is limited and is very motor specific. It lies typically between 0.05 and 30+ seconds (Refer to manufacturer's data sheet).

note The main contributor to the required peak/intermittent torque is acceleration, since torque equals inertia times acceleration. During constant speed, the motor needs only to maintain the constant torque due to friction forces.

Servo motor data sheets list typically a peak torque (either stall T_{PS} or rated T_{PR}) that is 2 to 3 times higher than the continuous torque (either stall T_{CS} or rated T_{CR}).

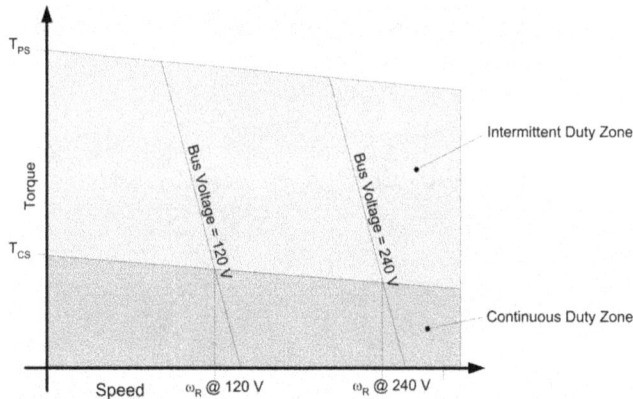

Picture 3.4.2.3.2: Influence of Bus Voltage on Torque/Speed

It is also necessary to know the voltage and current needed to operate the motor. Picture 3.4.2.3.2 demonstrates the influence of the bus voltage on the torque/speed curve. Higher voltage provides higher speed. The higher the torque requirements the higher will be the current. It is therefore important to consider the motor's torque constant k_t (k_t = Torque per Amp). Some manufacturers list a maximum speed, which is the speed at full voltage and no load. A more conservative, but more reliant approach is to list the rated continuous and intermittent torques which are the intersections of the peak (intermittent) and continuous torque curves with the rated speed (usually 3000+ rpm). The rated speed in turn is usually set at a point that covers a range of a somewhat constant torque level, until, at a certain speed the torque suddenly regresses.

In terms of comparing the application needs with the motor performance, the application's RMS torque and maximum speed must be verified with the motor's continuous torque/speed curve. The application's peak torque, which is a function of inertia and acceleration, must be compared to the intermittent torque/speed curve. In this case, the previously mentioned condition of limited time applies and must be taken into account.

The following picture shows an application whose torque and speed requirements are well within the motor's performance range.

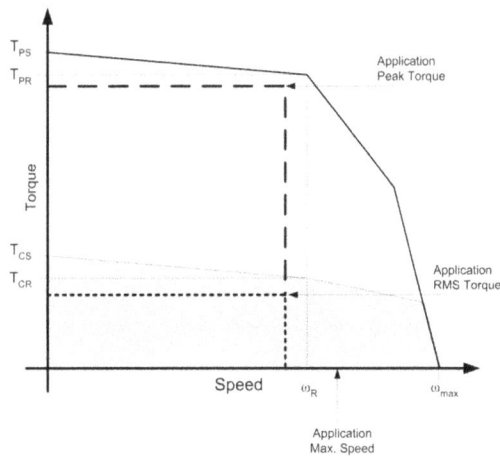

Picture **3.4.2.3.3:** Application 1 - OK

For reference purposes the next picture that shows an application whose torque and speed requirements are outside the reach of the motor's performance.

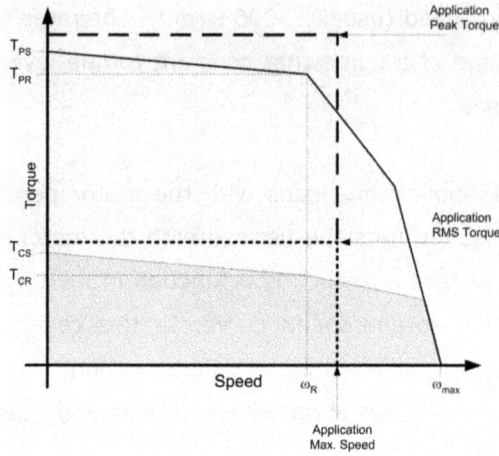

Picture 3.4.2.3.4: Application 2 - Failure

In certain cases it is possible to use a motor beyond the rated data, either torque or speed. The following example assumes a speed that is higher than the motor's rated speed, while the torque requirements are low.

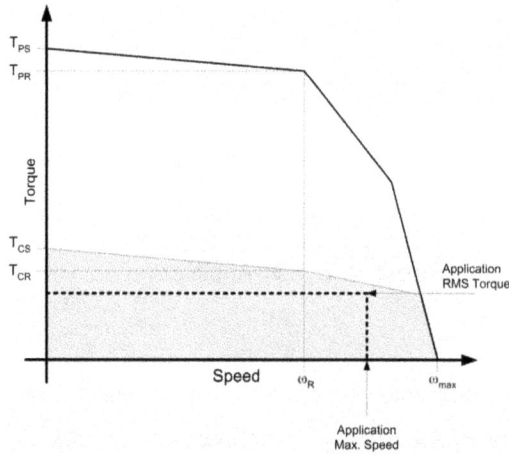

Picture 3.4.2.3.5: High speed, low RMS torque application

This example demonstrates an application speed that is beyond the motor's rated speed, however, with low enough required RMS torque to be covered by the continuous duty zone.

The next example shows a low speed, high RMS torque application.

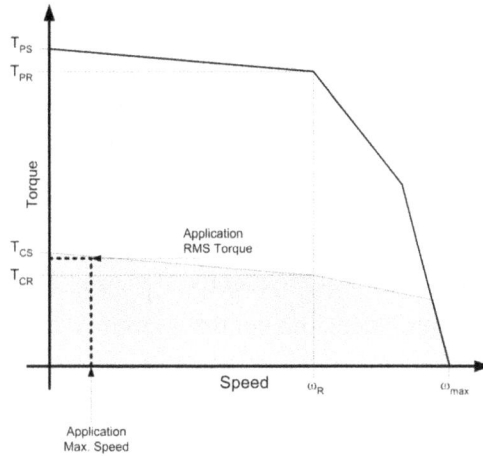

Picture 3.4.2.3.6: Low speed, high RMS torque application

This second example demonstrates an application RMS torque that is beyond the motor's rated continuous torque, however, with low enough speed requirements to be covered by the continuous duty zone.

3.4.2.4 Stepper Motor Performance Curves

The shape of stepper motor performance curves is usually in a totally different ballpark than servo motor curves. If their characteristics are misunderstood and they are applied based on the misconception, the application will fail. The costs attached to this failure may be significant.

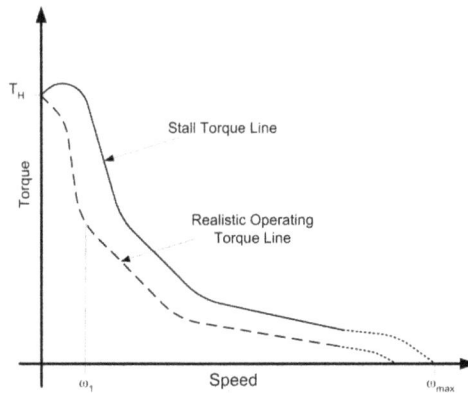

Picture 3.4.2.4.1: Typical Torque vs. Speed for Steppers

Just a brief look at the stepper motor performance curve in this picture reveals how difficult it can be to establish a distinctive rated torque or speed. Any settings would not do justice to the stepper's overall performance and thus the choice of a stepper motor must be very application specific. This example also emphasizes the importance of understanding and analyzing stepper motor performance curves. The "Stall Torque Line" in the picture represents the ideal performance as typically published by manufacturers of stepper motors and drive systems. This curve must be interpreted very differently than servo motor curves. Due to the open-loop nature of stepper systems and the complex dynamic interactions between motor, drive, load, and motion profile, a stepper motor will frequently stall well before reaching the ideal stall torque line. Unless feedback is provided, the control system will not be able to respond. Also, even the ideal torque declines rapidly above ω_1 (typically 100 – 600 rpm) to only 5 – 10% of holding torque T_H at ω_H (typically < 3000 rpm).

Thus, when selecting a stepper motor drive system, unless an application is extremely well defined and the loads do not change significantly, it is recommended that the designer makes use of a reduced torque speed curve similar to the "Realistic Operating Line" as shown in the previous picture. The realistic operating line is somewhat arbitrary defined as 50% of the stall torque line. The resulting selections will be much more robust and the application will usually be much more successful.

3.4.2.5   Servos vs. Steppers

Provided a stepper system will perform an application safely, it will generally be lower in costs than a similar servo system. The problem is defining a valid, consistent basis on which to compare them. The following picture attempts to provide some comparison criteria.

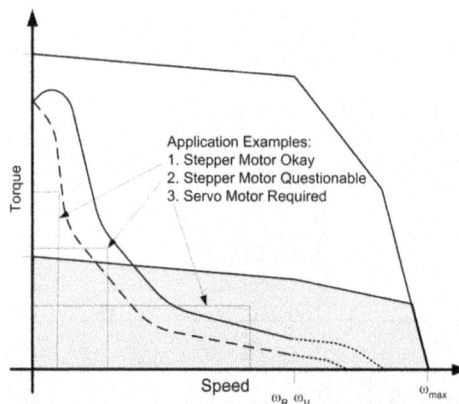

Picture 3.4.2.5.1: Servo and Stepper Comparison

The previous picture represents an overlay of the previously sample torque vs. speed curves for servo and stepper motors. Also shown are the torque and speed requirements of three different application examples. Note that the stepper motor's holding torque is over twice as much as the servo motor's rated torque. Also note that the "maximum" speed of the stepper motor is greater than the servo's rated speed.

It is also apparent that a selection based on the zero-speed torque values alone (which is very common) will lead to erroneous conclusions. Application 1 shows that a stepper motor would be a better choice for low speed applications requiring fairly high continuous and/or peak torque. Application 2 demonstrates that even at moderate speed a stepper motor may not have the torque to support the same application than a servo, even without applying the servo's intermittent torque. Application 3 is at higher speed and requires a servo motor, even though it requires less torque and the speed is below the stepper's speed abilities.

3.5 Special Design Considerations

In addition to the standard process of determining the system requirements it is necessary that the engineer is aware of the, let's say, "side effects" that the design might inflict on the mechanical system.

A gearbox, for instance, will almost automatically assure that the motor is smaller than without any gearing. Any gearing such as a gearbox or belt-pulley will reduce the torque required from the motor; hence, a smaller size motor will be sufficient. The downside (if it can be considered a disadvantage), is that a higher speed will be necessary. A holding brake, for instance, will add inertia to the load, but will reduce torque requirements during standstill in a vertical application.

3.5.1 Gearing

Any gearing such as a gearbox or belt-pulley will almost automatically assure that the motor is smaller than without any gearing. Gearing will reduce the torque required from the motor; hence, a smaller size motor will be sufficient. The downside (if it can be considered a disadvantage) is that a higher speed will be necessary.

$$T_{L \to M} = \frac{T_L}{N_r}$$

Equation 3.5.1.1: Effect of Transmission Ratio on Motor Torque[7]

Symbol	Description
$T_{L\text{->}M}$	Load torque reflected to motor
T_L	Load torque
N_r	Gearing transmission ratio

Equation 3.5.1.1 demonstrates the influence of the transmission ratio on the torque required by the motor, i.e. the higher the transmission ratio the lower will be the torque reflected to the motor. At the same time, the speed required from the motor will be multiplied by the transmission ratio.

$$v_M = v_L \times N_r$$

Equation 3.5.1.2: Effect of Transmission Ratio on Motor Speed

Symbol	Description
vM	Motor speed
vL	Load speed
Nr	Gearing transmission ratio

Gearing devices (i.e. devices where a transmission ratio affects the load inertia) are, for instance:

➤ Gearboxes
➤ Servo Reducers
➤ Belt-Pulleys
➤ Conveyors
➤ Rack-Pinions
➤ Leadscrews
➤ Nip Rolls

---

[7] This equation shows the basic relation between load and motor torque in a gearing application. In all consequence, the equation should also include the gearing device's efficiency. For more detailed information see chapter "4. Load Inertia and Torque Calculations."

In cases where a transmission ratio is not immediately available it needs to be calculated. In case of a leadscrew, use the pitch or lead. In case of a belt-pulley, conveyor and nip roll calculate the transmission rate as the ratio between motor side and load side diameters (See also the corresponding sections under paragraph "4. Load Inertia and Torque Calculation").

3.5.2 Holding Brake and Motor Torque Requirements

The effect of a holding brake on motor torque requirements is usually minimal, if not neglectable, in a regular rotary or linear horizontal motion application and is therefore not necessarily recommended. However, the torque and power reduction can be significant in case of vertical linear applications. The main purpose of a holding brake is to relief the motor from maintaining the holding torque during standstill periods in a vertical linear motion application. The use of a holding brake can save energy, not only due to the application of the brake during standstill, but also due to smaller motor size requirements.

The downside of using a holding brake, however, is that it adds inertia to the motor load, therefore increasing torque and power requirements during acceleration and deceleration. Some criteria to use a holding brake could be:

➢ Brake inertia is much smaller than the load inertia
➢ Duty cycle includes considerable standstill periods

The final decision to apply a holding brake should only be based on calculating the torque requirements – peak/intermittent as well as RMS torque - of all possible motor/brake combinations and then select the motor according to the lowest torque requirements. This is, of course, a time consuming and tedious process; but yet again a good example where motor sizing programs are a great help.

Example 1: Linear Horizontal Application without holding brake

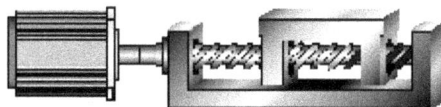

Picture 3.5.2.1: Leadscrew Application

Picture 3.5.2.1 shows a linear (leadscrew) application with no holding brake and we assume the following parameters:

Leadscrew Inertia	0.00000182	$in\text{-}lb\text{-}s^2$
Leadscrew Mounting Angle	0	degrees

Note that we (initially) chose a mounting angle of zero degrees to make this a linear horizontal application. Due to friction forces and effects of the mechanism efficiency there is a constant torque of roughly 0.0025 in-lb. The duty cycle was designed to have a forward movement and then have the leadscrew table return to the "home" position (as shown in the picture below). Back at the home position, we added a dwell time of 2 seconds.

The resulting torque requirements in reference to the duty cycle will look as follows[8]:

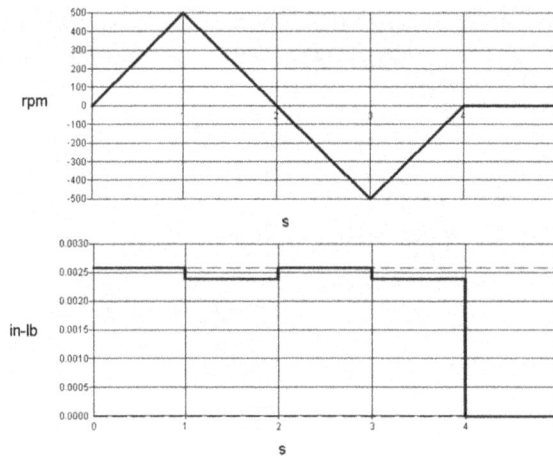

Picture 3.5.2.2: Velocity and Torque Graphs Horizontal Linear Application.

<u>Note:</u> The torque requirements as shown in picture 3.5.5.2 do not include the motor's inertia. Notice the required torque during acceleration and deceleration: Acceleration adds torque, deceleration reduces the torque.

The important part – since we intend to add a brake eventually – is the dwell time. As can be seen in picture 3.5.2.2 the total torque of the application will be zero during motor standstill, since this is a linear horizontal application, i.e. the motor does not need to maintain a holding torque. Ergo, in all consequence we do not need a holding brake. This picture will change quite a bit when we change the mounting angle to 90 degrees, i.e. creating a vertical application.

_____

Example 2: Linear Horizontal Application with holding brake

Picture 3.5.2.3: Leadscrew Application with brake

First, however, let's add a holding brake to the current linear horizontal application and observe the effects of the holding torque. We are assuming a brake inertia of 0.0001 in-lb-s^2, which is – for demonstration purposes - significantly higher than the leadscrew inertia.

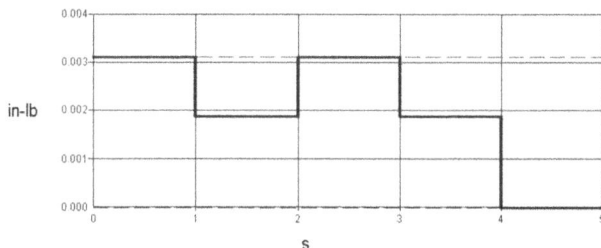

Picture 3.5.2.4: Effect of Holding Brake Inertia on Total Torque

As picture 3.5.2.4 clearly demonstrates the effect of the holding brake's inertia on the total torque, especially the intermittent torque, is quite dramatic in this example. The point really is that the use of a holding brake will increase the motor torque requirements to a certain degree.

In the next step let's set the leadscrew mounting angle to 90 % and make this a vertical application, however, <u>without</u> applying the holding brake at zero speed.

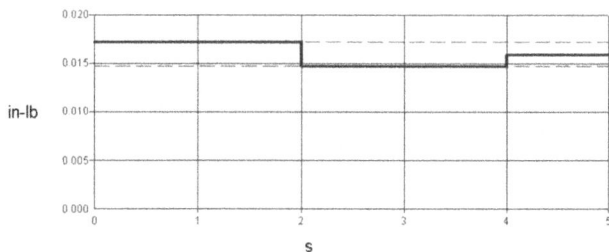

Picture 3.5.2.5: Total Torque in a Vertical Application

What happens here is that during acceleration the motor has to move the load upwards – against gravity -, while, during deceleration, it has to move the load downwards – with gravity.

In reference to picture 3.5.2.5, notice the current torque during dwell time, i.e. at standstill. The torque during dwell time is roughly 0.017 in-lb. The motor needs to maintain this torque in order to keep the leadscrew in the current position and to compensate for gravity forces.

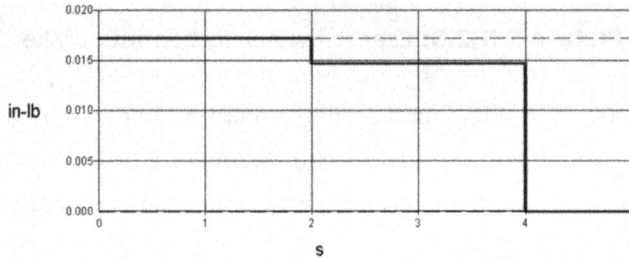

Picture 3.5.2.6: Total Torque with Holding Brake Applied

Notice yet again the torque during dwell time in picture 3.5.2.6, which is zero due to the applied holding brake. The maximum (intermittent) torque required from the motor will be in both cases, with or without a holding brake applied, the same, roughly over 0.017 in-lb. The main difference lies in the RMS torque, which is used to determine the motor's rated torque, i.e. the torque it can maintain constantly.

The equation to calculate the RMS torque is as shown here:

$$T_{RMS} = \sqrt{\frac{\left(T_a + T_c\right)^2 \times t_a + T_c^2 \times t_c + \left(T_d + T_c\right)^2 \times t_d + T_h^2 \times t_h}{t_a + t_c + t_d + t_h}}$$

Equation 3.5.2.1: RMS torque calculation

Symbol	Description
T_a	Acceleration torque
T_c	Constant torque
T_d	Deceleration torque
T_h	Holding torque
t_a	Acceleration time
t_c	Time of constant speed
t_d	Deceleration time
t_h	Holding (dwell) time

T_a, T_c, T_d and T_h as shown in the equation represent absolute values. Deceleration torque is usually negative. The motor, however, has to provide at least that amount of torque to drive the mechanical setup during deceleration.

According to equation 3.5.2.1 the RMS torque, according to the total torque as shown in picture 4 (brake not applied at zero speed) would be 0.016 in-lb, while the RMS torque according to picture 5 (brake applied at zero speed) would be 0.014 in-lb. The absolute difference of 0.002 in-lb may not seem much, but it is nevertheless a reduction of 12.5% in required motor torque. This 12.5% in savings can amount to significant energy savings over the lifetime of the mechanical system.

3.5.3 Vertical Applications

There are myriad types of linear vertical application where a load needs to move not only vertically (90°), but in any angle greater than zero. A zero angle would represent a linear horizontal application. However, true vertical (90°) applications are the most popular ones. Picture 3.5.3.1 shows a sample application, in this case a pick and place application.

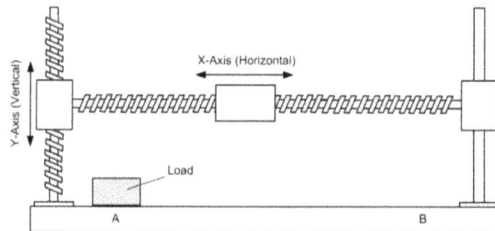

Picture 3.5.3.1: Pick & Place Application Example

This example uses lead screws for each axis. The lead screw supporting the vertical movement (Y-Axis) is used to pick up the load at position A and release it at position B. The lead screw supporting the horizontal movement (X-Axis) is used to move the load from A to B.

In this case, we focus on the Y-Axis, since it represents the vertical axis. The load of the Y-Axis consists of the X-Axis and, when moving the load from A to B, the actual load to be moved. The forces that are different compared to a linear horizontal application are the friction force and the gravity force. It is obvious that gravity will oppose any upward movement, but support any downward movement. Both forces, friction and gravity, will have an effect on the constant torque; it does not influence the acceleration torque.

The following equations for friction force, gravity force and constant torque show the influence of the application angle (γ).

$$F_{fr} = W_L \times \mu \times \cos(\gamma)$$

F_{fr} Friction Force
W_L Load Weight
μ Friction Coefficient
γ Theta (Application Angle)

Equation 3.5.3.1: Friction Force

$$F_G = W_L \times \sin(\gamma)$$

F_G Gravity Force
W_L Load Weight
γ Theta (Application Angle)

Equation 3.5.3.2: Gravity Force

$$T_C = \frac{F_{fr} + F_G}{2 \times \pi \times P \times \eta}$$

T_C Constant Torque
F_{fr} Friction Force
F_G Gravity Force
P Pitch
η Mechanism Efficiency

Equation 3.5.3.3: Constant Torque

3.5.4   Thrust Forces

Thrust forces are opposing or assisting forces on the load or weight of linear devices such as lead screws, actuators, conveyor belts, etc. Thrust forces will have an impact on the RMS and peak torque of the application and thus can have an essential influence on the motor selection. Thrust forces may not occur during the entire duty cycle, but only in certain segments; this is entirely application specific. It is important to determine where thrust forces occur and what values they have.

A good example to demonstrate the effect thrust forces is a drilling application: A drill moves by means of a lead screw. In terms of thrust force, it actually doesn't matter whether the movement is horizontal or vertical.

A description of the cycle of drilling a hole is as follows:

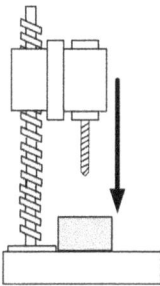

1. Move the drill towards the material surface at a specified (high) speed.

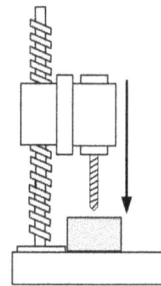

2. Slow down at a certain distance from the material's surface.

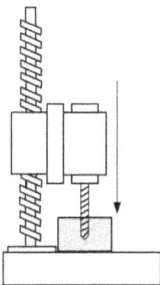

3. Continue moving the drill into the surface at a specified (slow) speed until you reach the desired drilling depth.

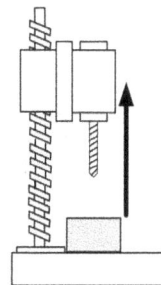

4. Retract the drill into the home position ("negative" speed).

Picture 3.5.4.1: Drilling Sample Application

The thrust forces apply as soon as the drill reaches the surface and continues to move into the material (step #3 as shown in picture 3.5.4.1), i.e. there will be forces opposing the linear movement of the drill and the lead screw, respectively.

The velocity graph according to the example will look like the following:

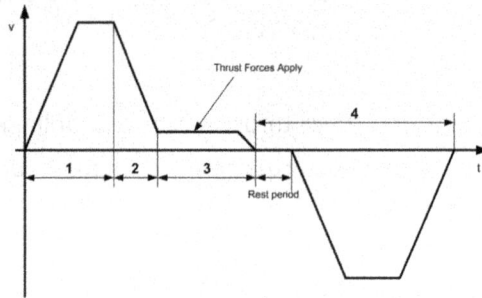

Picture 3.5.4.2: Drilling Duty Cycle

3.5.5 Load Variations

The load of a motion control application is not necessarily applied during the entire duty cycle, i.e. it is possible the load applies only during certain motion segments. The best example to demonstrate such a load is a pick-and-place application, i.e. a two axis system to support horizontal movement from x_1 to x_2 and a vertical movement from y_1 to y_2 as shown in the following picture.

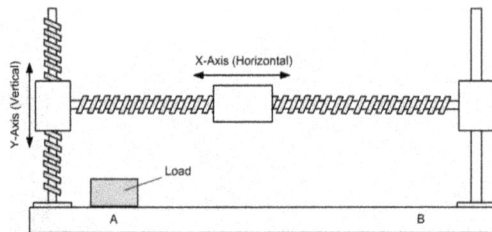

Picture 3.5.5.1: Pick & Place Application Example

This example uses lead screws for each axis. The lead screw supporting the vertical movement (Y-Axis) is used to pick up the load at position A and release it at position B. The lead screw supporting the horizontal movement (X-Axis) is used to move the load from A to B.

The duty cycle is as follows:

1. Y-Axis moves down to pick up the load
2. Y-Axis moves up to travel position; Load is now applied to axis Y
3. X-Axis moves from position A to B; Load is also applied to axis X

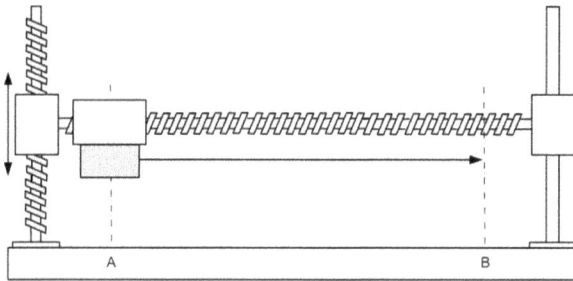

Picture 3.5.5.2: Pick & Place – Moving Load

1. Y-Axis moves down to release the load; Load is still applied
2. Y-Axis moves up to travel position; Load is not applied
3. X-Axis moves from position B to A; Load is not applied

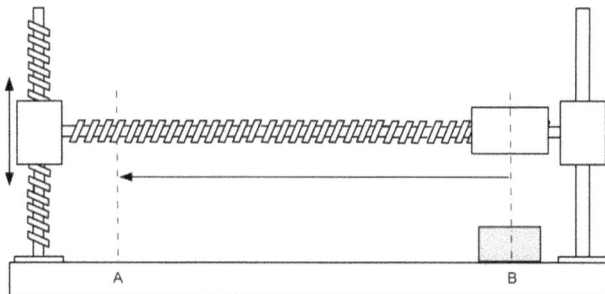

Picture 3.5.5.3: Pick & Place – Moving back to Pick Up Position

This type of application will affect the RMS and intermittent torque of the mechanical system and thus influence the size of both motors, the one for the X-Axis and the second for the Y-Axis. It is therefore important to consider changing loads when selecting a motor.

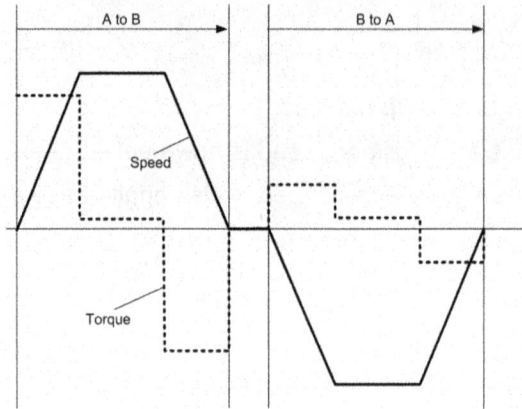

Picture 3.5.5.4: Pick & Place – Speed and Torque

Picture 3.5.5.4 demonstrates the duty cycle of the pick & place sample application (X-Axis) in terms of speed and torque requirements.

The torque requirements during the move from A to B (Load applied) will be higher than during the move from B to A (Load is not applied). The weight and inertia during the move from A to B includes the leadscrew (screw and table) plus the load, while during the move from B to A the weight and inertia includes only the leadscrew, hence the remaining lower torque in this section.

3.5.6 Multi-Dimensional (X-Y-Z) Applications

X-Y-Z applications are naturally not limited to three axes. Probably the most used multi-dimensional applications have two axes, for instance, pick & place and plotter applications. Robotics applications can have multiple dimensions, depending on the requirements.

The pick & place application has been used before to explain load variations and the effect of gravity forces on the constant torque. It is, yet again a good example to demonstrate a two-dimensional application.

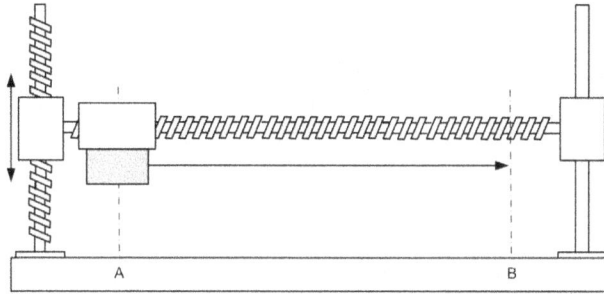

Picture 3.5.6.1: Pick & Place – Moving Load

The duty cycle is as follows:

1. Y-Axis moves down to pick up the load
2. Y-Axis moves up to travel position
3. X-Axis moves from position A to B
4. Y-Axis moves down to release the load
5. Y-Axis moves up to travel position
6. X-Axis moves from position B to A

The most important (and maybe really the only) consideration for multiple dimension applications in terms of motor sizing is that the load weight of an axis maybe the weight of another axis. In our example, the weight of the X-Axis is the load weight to the Y-Axis.

Table 3.5.6.1 demonstrates the load weights for an application with 5 axes:

Axis #	Load Weight
1 (Base axis)	Weight of axes 2, 3, 4 and 5 plus actual load weight
2 (First moving axis)	Weight of axes 3, 4 and 5 plus actual load weight
3 (Second moving axis)	Weight of axes 4 and 5 plus actual load weight
4 (Third moving axis)	Weight of axis 5 plus actual load weight
5 (Fourth moving axis)	Actual load weight

3.5.7 Thermal Considerations

As was mentioned in the very beginning of this book, the major costs of a motion control system are accumulated during its lifetime, primarily through power consumption. The lifetime of a motor, however, can be significantly shortened through thermal stress, i.e. running the motor at higher operating temperatures. Obviously, a 50°C environment is more demanding on the motor than room temperature (20° – 25°C).

Forced-air cooling can prolong the motor's life, but the best course to prevent an accelerated failure would be to determine the application's operational thermal profile and then specify the motor accordingly.

Picture 3.5.7.1 shows a sample performance curve (torque over speed) of a sample motor (Bayside K044200-KY2 at 300 Volt) at ambient temperatures of 20, 50 and 80°C.

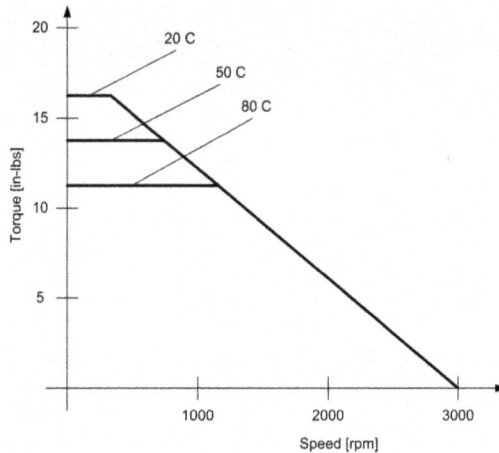

Picture 3.5.7.1: Thermal Influence on Motor Performance

This example demonstrates clearly the effect of ambient temperature on the motor's performance. In this case, it is apparent that the available torque decreases with higher ambient temperatures.

The temperature limits of a motor are primarily determined during the initial design considering factors such as (but not limited to):

➢ Class of insulating material
➢ Type of bearing grease
➢ Magnet temperature coefficient
➢ Motor housing style (totally enclosed or open)
➢ Thermal expansion of mechanical parts
➢ Heat transfer paths for convection or conduction

However, the motor design alone does not necessarily cover all application requirements; since the motor's operating conditions – which are a function of the application – affect the thermal profile. One way to prove whether or not the thermal stress can be harmful to the motor is by running lab tests and measure the temperature rise during operation. A more accurate and less time consuming approach is the calculation of the temperature rise.

Temperature-rise calculations are based on the motor's RMS current and its thermal resistance. Thermal resistance indicates how effectively the motor can dissipate the generated heat.

In order to calculate and verify the temperature rise the following parameters must be gathered from the motor data sheet:
1. Motor torque constant K_T in torque per amps (e.g. oz-in/A)
2. Hot motor-terminal resistance in Ohms
3. Max. running temperature in °C
4. Thermal resistance R_{TH}, measured in °C/W

The first step towards determining the temperature rise is the calculation of the RMS Torque as described in chapter "3.3.3 Load RMS torque". The RMS torque ("Root Mean Squared") represents the average torque over the entire duty cycle.

The motor's RMS current is calculated as follows:

$$I_{RMS} = \frac{T_{RMS}}{K_T}$$

I_{RMS}	RMS Current	Amps
T_{RMS}	RMS Torque	e.g. oz-in
K_T	Torque Constant	e.g. oz-in/A

Equation 3.5.7.1: RMS Current Calculation

The next step is to calculate the motor's power dissipation.

$$P_D = I_{RMS}^2 \times R_T$$

PD	Power Dissipation	W
I$_{RMS}$	RMS Current	Amps
R$_T$	Hot Motor-Terminal Resistance	Ohms

Equation 3.5.7.2: Power Dissipation Calculation

Finally, the temperature rise is calculated as follows:

$$\Delta T = P_D \times R_{TH}$$

ΔT	Temperature Rise	°C
PD	Power Dissipation	W
R$_{TH}$	Motor Thermal Resistance	°C/W

Equation 3.5.7.3: Temperature Rise Calculation

For the motor selection, the temperature rise must be added to the estimated ambient temperature and the result must be compared with the motor's maximum running temperature.

$$T_A + \Delta T \leq T_{Mmax}$$

TA	Ambient Temperature	°C
ΔT	Temperature Rise	°C
T$_{Mmax}$	Motor Max. Running Temperature	°C

Equation 3.5.7.4: Temperature Rise Evaluation

3.6   Sample application - comprised

According to the sample application used in previous chapters the motor requirements are:

- ➢ Max. Speed   1000 rpm
- ➢ Cont. Torque   0.164 in-lb (not including an oversize factor)
- ➢ Intermittent Torque   0.259 in-lb (not including an oversize factor)
- ➢ Rotor Inertia $0.000137...0.000822$ in-lb-sec^2

The inertia requirement ranges from a load/motor ratio of 1:1 to 6:1.

Sample Application:	Disk Application
	The motion objective of this sample application shall be to accomplish a trapezoidal motion profile, i.e. to accelerate the disk to a certain speed, hold the speed for some time and then decelerate to zero speed again. The disk shall be 2" in diameter, 1.2" in thickness. The material shall be steel, which has a material density of 0.28 lbs/in^3.
	Calculation of Maximum Load Acceleration
	The motion profile shall be of trapezoidal shape using the following parameters: $t_a = t_d = 1$ sec $t_c = 2$ sec $t_0 = 1$ sec $v_{max} = 1000$ rpm $= 16.67$ rev/sec The derived parameters are: $t_{total} = 4$ sec $a_{max} = v_{max} / t_a = 16.67 / 1 = 16.67$ rev/sec^2
	Determination of Maximum Load Speed
	As defined in the motion profile the maximum required speed is: $v_{max} = 1000$ rpm

	Calculation of Intermittent/Peak Torque
	Max. acceleration a_{max} = 16.67 rev/sec^2 as derived from the motion profile. The inertia of the disk is determined as: Disk Weight: $W = \pi \times \left(\dfrac{D}{2}\right)^2 \times L \times \rho$ Disk Inertia: $J_{disk} = \dfrac{W \times r^2}{2 \times g} = \dfrac{\pi \times L \times \rho \times r^4}{2 \times g}$ Using the disk dimensions as defined earlier the disk inertia will be: $$J_{disk} = 0.00137 in-lb-\sec^2$$ In the following, we assume a rotor inertia of 0.0011 in-lb-sec^2. $$J_{system} = 0.00137 + 0.0011 = 0.00247 in-lb-\sec^2$$ Since a disk does not inflict any friction, we assume the constant torque to be zero. As a result the acceleration torque represents the total torque: $$T_a = T_{total} = J_{system} \times \alpha \times 2 \times \pi = 0.259 in-lb$$

	Calculation of RMS Torque
	The equation to calculate the total RMS torque for a trapezoidal motion profile, including load and motor inertia, is: $$T_{RMS} = \sqrt{\dfrac{(T_a+T_c)^2 \times t_a + T_c^2 \times t_c + (T_d+T_c)^2 \times t_d + T_h^2 \times t_h}{t_a+t_c+t_d+t_h}}$$ By applying our current sample application data, i.e. assuming the constant torque to be zero and acceleration/deceleration times are 1 sec, the equation can be simplified to the following: $$T_{RMS} = \sqrt{\dfrac{T_a^2 + T_d^2}{t_a+t_c+t_d+t_h}} = \sqrt{\dfrac{(2 \times 0.259)^2}{5}} = 0.164 in-lb$$

Symbol	Description
W	Weight
Π	"PI" = 3.14159
D	Disk diameter
L	Disk thickness (length)
ρ	Material density
J_{disk}	Disk inertia
r	Disk radius
g	Gravity constant (386 in/sec^2)
T_a	Acceleration torque
T_c	Constant torque
T_d	Deceleration torque
T_h	Holding torque
t_a	Acceleration time
t_c	Time of constant speed
t_d	Deceleration time
t_h	Holding time

Load Inertia and Torque Calculation

Even though the calculation of complex mechanical devices seems to be very complicated at first glance, the math behind it is very much straight forward as long as there is an understanding that most components are derived from basic shapes (mostly a cylinder). Also helpful is the knowledge that speed transmissions are mostly accomplished using two cylinders (tooth wheels, pulleys, etc.) and that any transmission has an impact on acceleration, inertia and torque that is being reflected to the motor.

4.1 Basic Calculations

The motor selection is based on speed, inertia and torque comparisons. The calculation of mechanical devices is used to determine these parameters.

Even the most complex mechanical devices are calculated based on three "basic" components:

➢ Solid Cylinder

➢ Hollow Cylinder

➢ Solid Rectangular

The calculation of these components plus some fundamental equations are explained in the chapters to follow.

4.1.1   Fundamental Equations

1. Weight

$$W = V \times \rho$$

Weight (lb) = Volume (in3) x Material Density $\left(\dfrac{lb}{in^3} \right)$

2. Mass

$$m = \frac{W}{g\cdot} \quad \text{(Gravity Constant} \quad g = 386 \frac{in}{sec^2} = 9.80 \frac{m}{sec^2})$$

Mass $\left(\dfrac{lb - sec^2}{in} \right)$ = Weight (lb) / Gravity (in/sec²)

3. Radius

$$r = \frac{D}{2}$$

Radius (in) = Diameter (in) / 2

4. Total Torque

$$T_{total} = T_a + T_c$$

Total Torque (in-lb) = Acceleration Torque (in-lb) + Constant Torque (in-lb)

5. Acceleration Torque

$$T_a = J_{total} \times \alpha$$

Acceleration Torque (in-lb) = Mass Inertia (in-lb-sec2) x Angular Acceleration (radians/sec²)

6. Angular Acceleration

$$\alpha = \frac{\omega_{max}}{t_a} \times 2\pi$$

Angular Acceleration $\left(\dfrac{radians}{sec^2} \right)$ =

$$\dfrac{MaxSpeed}{AccelTime} \left(\dfrac{rev/sec}{sec} \right) \times \text{Rotational unit conversion} \left(\dfrac{2\pi \times radians}{rev} \right)$$

7. Constant Torque

$$T_C = \dfrac{F_{fr} + F_G}{2 \times \pi \times P \times \eta}$$

T_c = Torque due to forces such as friction, gravity, push-pull, preload, efficiency, etc.

8. RMS Torque

T_{RMS} = "Root Mean Squared" (average) torque over the entire duty cycle

$$T_{RMS} = \sqrt{\dfrac{\left(T_a + T_c\right)^2 \times t_a + T_c^2 \times t_c + \left(T_d + T_c\right)^2 \times t_d + T_h^2 \times t_h}{t_a + t_c + t_d + t_h}}$$

9. Friction Force

$$F_{fr} = \mu \times W_L$$

Friction Force (lb) = Friction Coefficient x Load Weight (lb)

4.1.2 Solid Cylinder

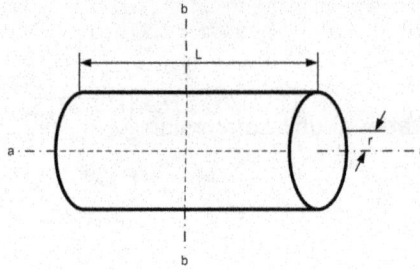

Picture 4.1.2.1: Solid Cylinder

$$A_{end} = \pi \times r^2$$

$$V = A_{end} \times L$$

$$J_{a-a} = \frac{m \times r^2}{2} = \frac{W \times r^2}{2 \times g} = \frac{\pi \times L \times \rho \times r^4}{2 \times g}$$

$$J_{b-b} = \frac{m}{12} \times \left(3r^2 + L^2\right)$$

A_{end}	Cylinder side area
g	Gravity Constant = 9.81 m-s^2 / 386 in-s^2
J_{a-a}	Inertia about axis a-a
J_{b-b}	Inertia about axis b-b
L	Cylinder Length
m	Mass
ρ	Mass density of material
r	Radius
V	Volume
W	Weight

Equation 4.1.2.1: Solid Cylinder

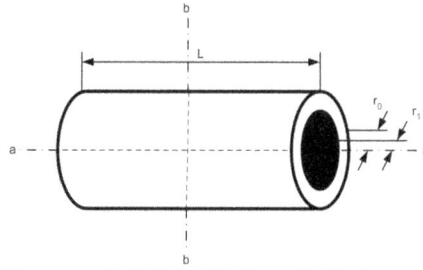

4.1.3 Hollow Cylinder

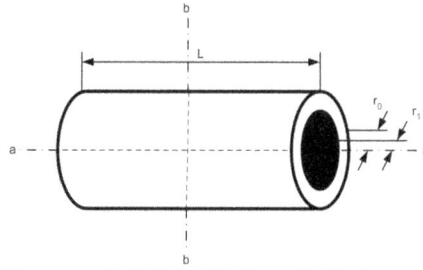

Picture 4.1.3.1: Hollow Cylinder

$$A_{end} = \pi \times \left(r_o^{\,2} - r_i^{\,2} \right)$$

$$V = A_{end} \times L$$

$$J_{a-a} = \frac{m}{2} \times (r_o^{\,2} + r_i^{\,2}) = \frac{W}{2g} \times (r_o^{\,2} + r_i^{\,2}) = \frac{\pi L \rho}{2g} \times \left(r_o^{\,4} - r_i^{\,4} \right)$$

$$J_{b-b} = \frac{m}{12} \times \left(3 r_o^{\,2} + 3 r_i^{\,2} + L^2 \right)$$

A_{end}	Cylinder end side area
g	Gravity Constant = 9.81 m-s^2 / 386 in-s^2
J_{a-a}	Inertia about axis a-a
J_{b-b}	Inertia about axis b-b
L	Cylinder Length
m	Mass
ρ	Mass density of material
r_o	Outer Radius
r_i	Inner Radius
V	Volume
W	Weight

Equation 4.1.3.1: Hollow Cylinder

4.1.4   Rectangular Block

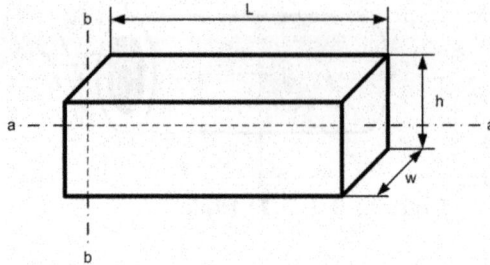

Picture 4.1.4.1: Rectangular Block

$$A_{end} = h \times w \qquad A_{side} = L \times h \qquad V = L \times h \times w$$

$$J_{a-a} = \frac{m}{12} \times \left(h^2 + w^2 \right)$$

$$J_{b-b} = \frac{m}{12} \times \left(4L^2 + w^2 \right) \quad \text{- if short}$$

$$J_{b-b} = \frac{m}{3} \times L^2 \quad \text{- if h, w << L}$$

Aend	Block end area
Aside	Block side area
h	Block height
Ja-a	Inertia about axis a-a
Jb-b	Inertia about axis b-b
L	Block length
m	Mass
V	Volume of block
w	Block width

Equation 4.1.4.1: Rectangular Block

4.2   Calculation of Mechanical Components

This chapter describes the calculation of more complex mechanisms. While the actual math behind these calculations is pretty much straight forward, it is nevertheless helpful to be aware of a few circumstances. All mechanical devices that will be discussed in the following support either a rotary or a linear motion. Some of the rotary devices, such as a gearbox, are only used for speed transmission (affecting the speed needed from the motor) and others, such as a rotary table are actual loads. All linear devices are also loads, i.e. they do the actual work as defined by the engineer. Mechanical devices can be categorized as follows.

Rotary Loads

- ➢ Nip Roll
- ➢ Rotary Table
- ➢ Winder

Linear Loads

For the calculation of linear devices, it is necessary to know the lead or pitch, which represents the linear length traveled during one revolution at the input shaft. The lead or pitch can also be calculated, for instance, for a conveyor, by looking at the circumference of the motor side pulley.

- ➢ Conveyor
- ➢ Linear Actuator
- ➢ Leadscrew
- ➢ Rack-Pinion

Speed Transmissions

For the calculation of speed transmission devices, it is necessary to know the transmission ratio. In case of a timing belt, the ratio can be determined as the ratio between motor side pulley diameter and load side pulley diameter.

- ➢ Belt-Pulley (Timing Belt)
- ➢ Gearbox (Servo Reducer)

Miscellaneous (all rotary)

- ➢ Brake
- ➢ Coupling
- ➢ Disk

There are also two important parameters needed for the calculation of mechanical devices; they can, however, cause a lot of confusion, either due to lack of knowledge or simply that the data is not available from the manufacturer of the device.[9]

These parameters are:

1. **Friction Coefficient**
2. **Mechanism Efficiency**

Both parameters are also very important for linear vertical application, since they do influence the constant torque differently during upwards and downwards movements.

The coefficient of friction is a dimensionless quantity and is being used to calculate the friction force. In case of the mechanical devices as discussed in the following chapter, the friction force describes the friction between two surfaces. For instance, in case of a leadscrew, the friction force applies between the screw and the nut. In most cases, it is sufficient to use a coefficient of about 0.15, which represents steel on steel (greased).

For further examples of friction, coefficient data refer to "Appendix F – Friction Coefficients".

The mechanism efficiency is also a dimensionless quantity between 0 and 1.0 or 0 to 100%. It describes the losses, for instance, between two tooth wheels in a gearbox or between the screw and nut of a leadscrew. For example, the efficiency of a ball-screw can be between ~0.85 and 0.95.

For further examples of mechanism efficiencies data refer to "Appendix E – Mechanism Efficiencies".

Hear Ye! Hear Ye!

Please be aware that some mechanical devices such as leadscrews, actuators, servo reducers, etc. can only be used within certain speed and torque ranges. Make sure to study the corresponding data sheets and make sure that these devices are not being operated beyond their terminal speed and torque capabilities.

---

[9] Some manufacturers of, for instance, linear actuators underestimate the importance of friction coefficient and efficiency for the purpose of motor sizing. These data are not necessarily included in their data sheets.

4.2.1   Disk

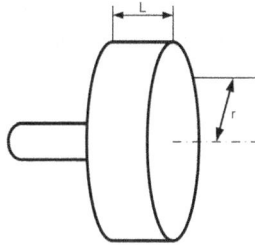

Picture 4.2.1.1: Disk

The disk is generally the most used mechanical component when it comes to inertia calculations. A disk can be used to resemble other mechanical components such as pulleys, screws, pinions, gears, couplings, etc. Naturally, the disk is nothing else but a solid (or hollow) cylinder and thus the equations for a disk and solid (hollow) cylinder are the same.

$$J_{Disk} = \frac{m \times r^2}{2} = \frac{W \times r^2}{2 \times g} = \frac{\pi \times L \times \rho \times r^4}{2 \times g}$$

g	Gravity Constant = 9.81 m-s^2 / 386 in-s^2
J_{Disk}	Disk Inertia
L	Disk Length/Thickness
m	Mass
ρ	Mass density of material
r	Radius
W	Weight

Equation 4.2.1.1: Disk

4.2.2   Chain Drive

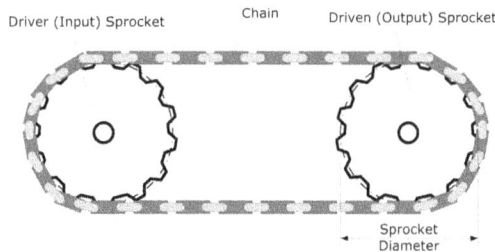

Picture 4.2.2.1: Chain-Drive

The chain drive consists of two tooth wheels and a chain. In the following, the disk on the motor side is called the driver sprocket; the disk on the load side is called the driven Sprocket. Both sprockets are calculated like a regular disk (cylinder) or, if necessary, a hollow cylinder.

$$J_{Disk} = \frac{m \times r^2}{2} = \frac{W \times r^2}{2 \times g} = \frac{\pi \times L \times \rho \times r^4}{2 \times g}$$

g	Gravity Constant = 9.81 m-s² = 386 in-s²
J_{Disk}	Disk Inertia
L	Disk Length/Thickness
m	Mass
ρ	Mass density of material
r	Radius
W	Weight

Equation 4.2.2.1: Sprocket Inertia

The transmission rate of the chain sprocket device is derived from either the sprocket diameters or the number of teeth on both sprockets.

$$N_r = \frac{D_{SL}}{D_{SM}} = \frac{N_{TL}}{N_{TM}}$$

N_r	Transmission ratio
D_{SL}	Diameter of sprocket on load side
D_{SM}	Diameter of sprocket on motor side
N_{TL}	Number of teeth on load side
N_{TM}	Number of teeth on motor side

Equation 4.2.2.2: Chain-Sprocket transmission ratio

The total inertia of the chain sprocket device includes the driver sprocket inertia, the chain inertia reflected to the motor side and the inertia of the driven sprocket reflected to the motor side.

$$J_{total} = J_{SM} + J_{SL \to M} + J_{C \to M} + J_{L \to M}$$

$$J_{SL \to M} = \left(\frac{1}{N_r}\right)^2 \times \frac{J_{SL}}{e} \qquad J_{C \to M} = \frac{W_C}{g \times e} \times \left(\frac{D_{SM}}{2}\right)^2 \qquad J_{L \to M} = \left(\frac{1}{N_r}\right)^2 \times \frac{J_L}{e}$$

e	Mechanism efficiency
g	Gravity Constant = 9.81 m-s2 = 386 in-s2
DSM	Diameter of sprocket on motor side
Jtotal	Total inertia including load
JSM	Inertia of sprocket on the motor side
JSL->M	Load side sprocket inertia reflected to motor
JC->M	Chain inertia reflected to motor
JL->M	Load inertia reflected to motor
JSL	Load side sprocket inertia
JL	Load inertia
Nr	Transmission ratio
WC	Chain weight

Equation 4.2.2.3: Chain-Sprocket inertia calculation

$$T_{L \to M} = \frac{T_L}{N_r \times e}$$

e	Mechanism Efficiency
N_r	Transmission ratio
$T_{L->M}$	Load torque reflected to motor
T_L	Load torque

Equation 4.2.2.4: Chain-Sprocket load torque calculation

4.2.3 Coupling

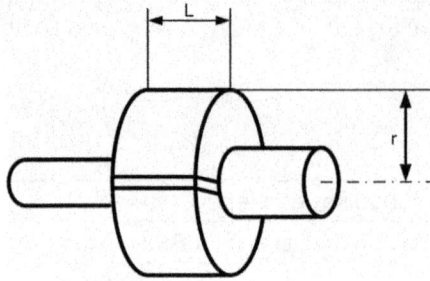

Picture 4.2.3.1: Coupling

The coupling is nothing else but a solid (or hollow) cylinder and thus the equations for a coupling and solid (hollow) cylinder are the same. Usually the coupling is calculated like a hollow cylinder.

$$A_{end} = \pi \times \left(r_o^2 - r_i^2\right) \qquad V = A_{end} \times L$$

$$J_C = \frac{m}{2} \times (r_o^2 + r_i^2) = \frac{W}{2g} \times (r_o^2 + r_i^2) = \frac{\pi L \rho}{2g} \times \left(r_o^4 - r_i^4\right)$$

A_{end}	Cylinder end side area
g	Gravity Constant = 9.81 m-s^2 / 386 in-s^2
J_C	Coupling inertia
L	Coupling Length
m	Mass
ρ	Mass density of material
r_o	Outer Radius
r_i	Inner Radius
V	Volume
W	Weight

Equation 4.2.3.1: Coupling Inertia

4.2.4   Gears

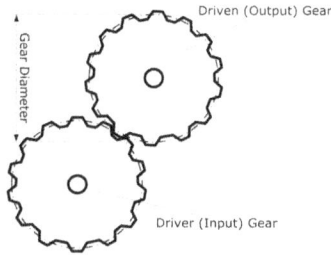

Picture 4.2.4.1: Gears

Gears are basically two tooth wheels where one wheel (motor side) drives the other one (load side). Both wheels are calculated like a regular disk (cylinder) or, if necessary, a hollow cylinder.

$$J_{Disk} = \frac{m \times r^2}{2} = \frac{W \times r^2}{2 \times g} = \frac{\pi \times L \times \rho \times r^4}{2 \times g}$$

g	Gravity Constant = 9.81 m-s^2 = 386 in-s^2
J_{Disk}	Disk Inertia
L	Disk Length/Thickness
m	Mass
ρ	Mass density of material
r	Radius
W	Weight

Equation 4.2.4.1: Gear Inertia

The transmission rate of the gear device is derived from either the gear diameters or the number of teeth on both gears.

$$N_r = \frac{D_{GL}}{D_{GM}} = \frac{N_{TL}}{N_{TM}}$$

N_r	Transmission ratio
D_{GL}	Diameter of gear on load side
D_{GM}	Diameter of gear on motor side
N_{TL}	Number of teeth on load side
N_{TM}	Number of teeth on motor side

Equation 4.2.4.2: Gears transmission ratio

The total inertia of the gear device includes the driver gear inertia and the inertia of the driven gear reflected to the motor side.

$$J_{total} = J_{GM} + J_{GL \to M} + J_{L \to M}$$

$$J_{GL \to M} = \left(\frac{1}{N_r} \right)^2 \times \frac{J_{GL}}{e} \qquad J_{L \to M} = \left(\frac{1}{N_r} \right)^2 \times \frac{J_L}{e}$$

e	Mechanism efficiency
g	Gravity Constant = 9.81 m-s^2 = 386 in-s^2
J_{total}	Total inertia including load
J_{GM}	Inertia of gear on the motor side
$J_{GL->M}$	Load side gear inertia reflected to motor
$J_{L->M}$	Load inertia reflected to motor
J_{GL}	Load side gear inertia
J_L	Load inertia
N_r	Transmission ratio

Equation 4.2.4.3: Gears inertia calculation

$$T_{L \to M} = \frac{T_L}{N_r \times e}$$

e	Mechanism Efficiency
N_r	Transmission ratio
$T_{L->M}$	Load torque reflected to motor
T_L	Load torque

Equation 4.2.4.4: Gears load torque calculation

4.2.5   Gearbox / Servo Reducer

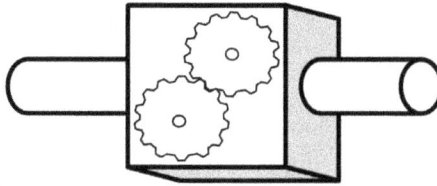

Picture 4.2.5.1: Gearbox

A gearbox and the previously documented gears are virtually identical in their functionality, but in this case, we are not concerned about the inner works of the gearbox. All that is really important to know is the gearbox's inertia, the transmission rate and efficiency plus the inertia of the load. The math is nevertheless basically the same.

Servo reducers are essentially identical to gearboxes, but some manufacturers prefer to use a more sophisticated term than gearbox, because their inner mechanical construction can be far more complex than a simple gearbox constructed through regular tooth wheels. While the math is the same as for gearboxes, servo reducers do have limitations in regards to input/output speed and torque.

$$J_{total} = J_G + J_{L \to M}$$

$$J_{L \to M} = \left(\frac{1}{N_r} \right)^2 \times \frac{J_L}{e}$$

e	Mechanism efficiency
J_{total}	Total inertia including load
J_G	Gearbox inertia
$J_{L->M}$	Load inertia reflected to motor
J_L	Load inertia
N_r	Transmission ratio

Equation 4.2.5.1: Gearbox inertia calculation

$$T_{L \to M} = \frac{T_L}{N_r \times e}$$

e	Mechanism Efficiency
N_r	Transmission ratio
$T_{L->M}$	Load torque reflected to motor
T_L	Load torque

Equation 4.2.5.2: Gearbox load torque calculation

4.2.6 Belt-Pulley

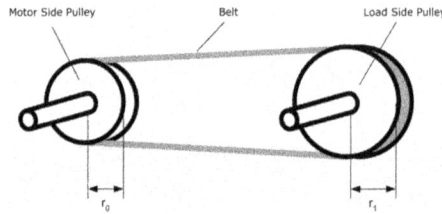

Picture 4.2.6.1: Belt Pulley

A Belt Pulley (or timing belt) consists of two disks (pulleys) and a belt. The disk on the motor side is called the motor side pulley; the disk on the load side is called the load side pulley. Both pulleys are calculated like a regular disk (cylinder) or, if necessary, a hollow cylinder.

$$J_{Disk} = \frac{m \times r^2}{2} = \frac{W \times r^2}{2 \times g} = \frac{\pi \times L \times \rho \times r^4}{2 \times g}$$

g	Gravity Constant = 9.81 m-s^2 = 386 in-s^2
J_{Disk}	Disk Inertia
L	Disk Length/Thickness
m	Mass
ρ	Mass density of material
r	Radius
W	Weight

Equation 4.2.6.1: Pulley Inertia

The transmission rate of the belt pulley is derived from the pulley diameters.

$$N_r = \frac{D_{PL}}{D_{PM}}$$

N_r	Transmission ratio
D_{PL}	Diameter of load side pulley
D_{PM}	Diameter of motor side pulley

Equation 4.2.6.2: Belt Pulley transmission ratio

The total inertia of the belt pulley includes the motor side pulley inertia, the belt inertia reflected to the motor side and the inertia of the load side pulley reflected to the motor side.

$$J_{total} = J_{PM} + J_{PL \to M} + J_{B \to M} + J_{L \to M}$$

$$J_{PL \to M} = \left(\frac{1}{N_r}\right)^2 \times \frac{J_{PL}}{e} \qquad J_{B \to M} = \frac{W_B}{g \times e} \times \left(\frac{D_{PM}}{2}\right)^2 \qquad J_{L \to M} = \left(\frac{1}{N_r}\right)^2 \times \frac{J_L}{e}$$

e	Mechanism efficiency
g	Gravity Constant = 9.81 m-s2 = 386 in-s2
DPM	Diameter of motor side pulley
Jtotal	Total inertia including load
JPM	Inertia of motor side pulley
JPL->M	Load side pulley inertia reflected to motor
JB->M	Belt inertia reflected to motor
JL->M	Load inertia reflected to motor
JPL	Load side pulley inertia
JL	Load inertia
Nr	Transmission ratio
WB	Belt weight

Equation 4.2.6.3: Belt Pulley inertia calculation

$$T_{L \to M} = \frac{T_L}{N_r \times e}$$

e	Mechanism Efficiency
N_r	Transmission ratio
$T_{L\text{->}M}$	Load torque reflected to motor
T_L	Load torque

Equation 4.2.6.4: Belt Pulley load torque calculation

4.2.7 Conveyor

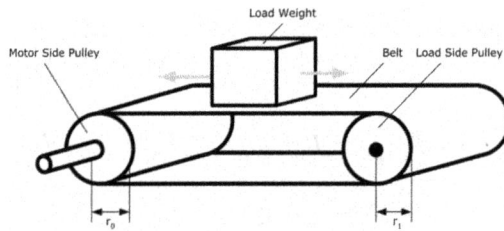

Picture 4.2.7.1: Conveyor

A Conveyor consists of two disks (pulleys) and a belt. The disk on the motor side is called the motor side pulley; the disk on the load side is called the load side pulley. Both pulleys are calculated like a regular disk (cylinder) or, if necessary, a hollow cylinder.

$$J_{Disk} = \frac{m \times r^2}{2} = \frac{W \times r^2}{2 \times g} = \frac{\pi \times L \times \rho \times r^4}{2 \times g}$$

g	Gravity Constant = 9.81 m-s^2 = 386 in-s^2
J_{Disk}	Disk Inertia
L	Disk Length/Thickness
m	Mass
ρ	Mass density of material
r	Radius
W	Weight

Equation 4.2.7.1: Pulley Inertia

The transmission rate of the conveyor is derived from the pulley diameters.

$$N_r = \frac{D_{PL}}{D_{PM}}$$

N_r	Transmission ratio
D_{PL}	Diameter of load side pulley
D_{PM}	Diameter of motor side pulley

Equation 4.2.7.2: Conveyor transmission ratio

The total inertia of the conveyor includes the motor side pulley inertia, the belt inertia reflected to the motor side and the inertia of the load side pulley reflected to the motor side.

$$J_{total} = J_{PM} + J_{PL \to M} + J_{B \to M} + J_{L \to M}$$

$$J_{PL \to M} = \left(\frac{1}{N_r}\right)^2 \times \frac{J_{PL}}{e} \quad J_{B \to M} = \frac{W_B}{g \times e} \times \left(\frac{D_{PM}}{2}\right)^2 \quad J_{L \to M} = \left(\frac{1}{N_r}\right)^2 \times \frac{J_L}{e}$$

e	Mechanism efficiency
g	Gravity Constant = 9.81 m-s^2 = 386 in-s^2
D_{PM}	Diameter of motor side pulley
J_{total}	Total inertia including load
J_{PM}	Inertia of motor side pulley
$J_{PL->M}$	Load side pulley inertia reflected to motor
$J_{B->M}$	Belt inertia reflected to motor
$J_{L->M}$	Load inertia reflected to motor
J_{PL}	Load side pulley inertia
J_L	Load inertia
N_r	Transmission ratio
W_B	Belt weight

Equation 4.2.7.3: Conveyor inertia calculation

$$T_{L \to M} = \frac{T_L}{N_r \times e}$$

e	Mechanism Efficiency
N_r	Transmission ratio
$T_{L->M}$	Load torque reflected to motor
T_L	Load torque

Equation 4.2.7.4: Conveyor load torque calculation

4.2.8 Leadscrew

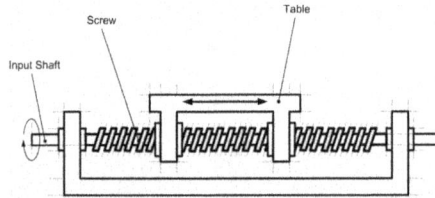

Picture 4.2.8.1: Leadscrew

In order to calculate leadscrew inertia the screw is considered a disk, thus the inertia of the screw is calculated like a disk. A leadscrew's transmission ratio is called a pitch (rev/in) or lead (in/rev). The total inertia of a leadscrew includes the reflected load inertia (in this case the load contains of the table weight and the actual load weight) and the screw inertia.

$$J_{Disk} = \frac{m \times r^2}{2} = \frac{W \times r^2}{2 \times g} = \frac{\pi \times L \times \rho \times r^4}{2 \times g}$$

g	Gravity Constant = 9.81 m-s^2 = 386 in-s^2
J_{Disk}	Disk Inertia
L	Disk Length/Thickness
m	Mass
ρ	Mass density of material
r	Radius
W	Weight

Equation 4.2.8.1: Screw Inertia

$$J_{total} = J_S + J_{L \to M}$$

$$J_{L \to M} = \frac{W_L + W_T}{g \times e} \times \left(\frac{1}{2\pi \times P_S} \right)^2$$

e	Mechanism efficiency
g	Gravity Constant = 9.81 m-s^2 = 386 in-s^2
J$_{total}$	Total inertia including load
J$_S$	Screw inertia
J$_{L->M}$	Load inertia reflected to motor
P$_S$	Leadscrew pitch
W$_L$	Load weight
W$_T$	Table weight

Equation 4.2.8.2: Leadscrew inertia calculation

$$F_g = \left(W_L + W_T \right) \times \sin \gamma \qquad F_{fr} = \mu \times \left(W_L + W_T \right) \times \cos \gamma$$

$$T_{L \to M} = \left(\frac{F_P + F_g + F_{fr}}{2\pi \times P_S \times e} \right) + T_P$$

e	Mechanism efficiency
γ	Application angle (Theta)
µ	Friction Coefficient[10]
F$_g$	Gravity Forces
F$_{fr}$	Friction forces
F$_P$	Push pull forces
P$_S$	Leadscrew pitch
T$_{L->M}$	Load torque reflected to motor
T$_P$	Preload torque
W$_L$	Load weight
W$_T$	Table weight

Equation 4.2.8.3: Leadscrew torque calculation

---

[10] For some examples of a friction coefficient, see Appendix F.

4.2.9 Linear Actuator[11]

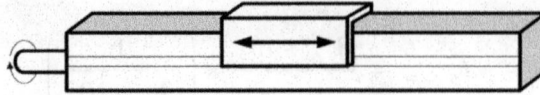

Picture 4.2.9.1: Linear Actuator

A linear actuator is not really a mechanical device on its own; the inner workings can be a leadscrew or a belt-pulley (timing belt). The following takes a very general approach to the calculation of a linear actuator. The calculation of linear actuators is usually very manufacturer specific. The inertia, for instance, depends very much on the length of the actuator, the type of slide and other manufacturer specific equipment. Please refer to the manufacturers' data sheets for detailed information on inertia calculation.

$$J_{total} = J_A + J_{L \to M}$$

$$J_{L \to M} = \frac{W_L + W_T}{g \times e} \times \left(\frac{1}{2\pi \times P_S} \right)^2$$

e	Mechanism efficiency
g	Gravity Constant = 9.81 m-s^2 = 386 in-s^2
J$_{total}$	Total inertia including load
J$_A$	Actuator inertia
J$_{L->M}$	Load inertia reflected to motor
P$_S$	Actuator pitch
W$_L$	Load weight
W$_T$	Table (slide) weight

Equation 4.2.9.1: Actuator inertia calculation

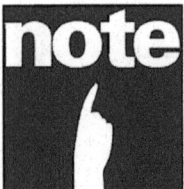

note

Manufacturers of linear actuators usually provide an "inertia factor" to calculate the load inertia reflected to the motor. In this case the reflected inertia is calculated as:

Reflected Inertia = Load Weight x Inertia Factor

---

[11] The calculation of linear actuators can be very manufacturer-specific and all specifics would provide enough material for another book.

$$F_g = \left(W_L + W_T\right) \times \sin\gamma \qquad F_{fr} = \mu \times \left(W_L + W_T\right) \times \cos\gamma$$

$$T_{L \to M} = \left(\frac{F_P + F_g + F_{fr}}{2\pi \times P_S \times e}\right) + T_P$$

e	Mechanism efficiency
γ	Application angle (Theta)
μ	Friction Coefficient[12]
Fg	Gravity Forces
Ffr	Friction forces
FP	Push pull forces
PS	Actuator pitch
TL->M	Load torque reflected to motor
TP	Preload torque
WL	Load weight
WT	Table weight

Equation 4.2.9.2: Actuator torque calculation

note

Manufacturers of linear actuators do not necessarily provide all data needed for the equations as listed in this chapter, even though parameters like, for instance friction coefficient, can have an impact on the total torque especially in vertical applications. For precise calculation results, please refer to the manufacturers' specifications.

4.2.10 Nip Roll

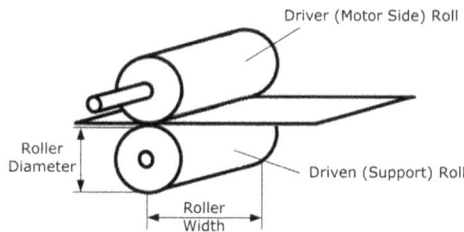

Picture 4.2.10.1: Nip Roll

---

[12] For some examples of a friction coefficient, see Appendix F.

A nip roll consists basically of two rollers where one roller (motor side) drives the other one (load side, support roller). Both rollers are calculated like a regular disk (cylinder) or, if necessary, a hollow cylinder.

$$J_{Disk} = \frac{m \times r^2}{2} = \frac{W \times r^2}{2 \times g} = \frac{\pi \times L \times \rho \times r^4}{2 \times g}$$

g	Gravity Constant = 9.81 m-s2 = 386 in-s2
JDisk	Disk Inertia
L	Disk Length/Thickness
m	Mass
ρ	Mass density of material
r	Radius
W	Weight

Equation 4.2.10.1: Roller Inertia

The nip roll is a load device, i.e. there is no other load attached to the nip roll. For that reason the transmission ratio is not significant for speed calculation, but it is needed to calculate the driven rollers inertia that is reflected to the motor. The transmission rate of the nip roll is derived from the roller diameters.

$$N_r = \frac{D_{RL}}{D_{RM}}$$

N_r	Transmission ratio
D_{RL}	Diameter of roller on load side
D_{RM}	Diameter of roller on motor side

Equation 4.2.10.2: Nip roll transmission ratio

The total inertia of the nip roll includes the driver roller inertia and the inertia of the driven roller reflected to the motor side.

$$J_{total} = J_{RM} + J_{RL \to M}$$

$$J_{RL \to M} = \left(\frac{1}{N_r} \right)^2 \times \frac{J_{RL}}{e}$$

e	Mechanism efficiency
g	Gravity Constant = 9.81 m-s2 = 386 in-s2
Jtotal	Total inertia
JRM	Inertia of roller on the motor side
JRL->M	Load side roller inertia reflected to motor
JRL	Load side roller inertia
Nr	Transmission ratio

Equation 4.2.10.3: Nip roll inertia calculation

$$T_{L \to M} = \frac{T_L}{N_r \times e}$$

e	Mechanism Efficiency
Nr	Transmission ratio
TL->M	Load side torque reflected to motor
TL	Load side torque

Equation 4.2.10.4: Nip roll load torque calculation

One parameter that should not be neglected on a nip roll is the web tension, which should be added as a constant torque of the motor side roller.

4.2.11 Rack Pinion

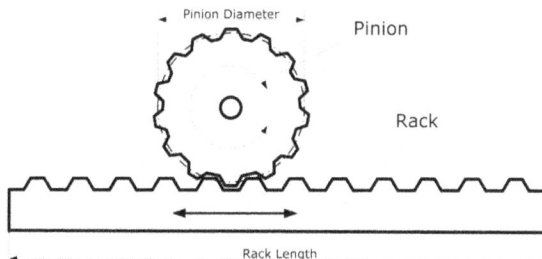

Picture 4.2.11.1: Rack Pinion

A rack pinion mechanism consists of a rotating disk (pinion) and a rectangular block (rack). The pinion inertia can be calculated using the equations of a disk.

$$J_{Disk} = \frac{m \times r^2}{2} = \frac{W \times r^2}{2 \times g} = \frac{\pi \times L \times \rho \times r^4}{2 \times g}$$

g	Gravity Constant = 9.81 m-s^2 = 386 in-s^2
J_{Disk}	Disk Inertia
L	Disk Length/Thickness
m	Mass
ρ	Mass density of material
r	Radius
W	Weight

Equation 4.2.11.1: Pinion Inertia

Since the rack is being used for linear (not rotary) motion we are only interested in its weight (See equation 4.2.11.2), which will then be used to calculate the inertia (See equation 4.2.11.3).

$$V = L \times h \times w$$
$$W = V \times \rho$$

h	Block height
L	Block length
m	Mass
ρ	Material density[13]
V	Volume of block
w	Block width

Equation 4.2.11.2: Rack weight

---

[13] For examples of material densities, see Appendix D.

$$J_{total} = J_P + J_{L \to M}$$

$$J_{L \to M} = \frac{W_L + W_R}{g \times e} \times \left(\frac{D_P}{2}\right)^2$$

e	Mechanism efficiency
g	Gravity Constant = 9.81 m-s^2 = 386 in-s^2
D_P	Pinion diameter
J_{total}	Total inertia including load
J_P	Pinion inertia
$J_{L->M}$	Load inertia reflected to motor
W_L	Load weight
W_R	Rack weight

Equation 4.2.11.3: Rack Pinion inertia calculation

$$F_g = \left(W_L + W_T\right) \times \sin \gamma \quad F_{fr} = \mu \times \left(W_L + W_T\right) \times \cos \gamma$$

$$T_{L \to M} = \left(\frac{F_P + F_g + F_f r}{e}\right) \times \left(\frac{D_P}{2}\right)$$

D_P	Pinion Diameter
e	Mechanism efficiency
γ	Application angle (Theta)
μ	Friction Coefficient[14]
F_g	Gravity Forces
F_{fr}	Friction forces
F_P	Push pull forces
P_S	Leadscrew pitch
$T_{L->M}$	Load torque reflected to motor
W_L	Load weight
W_T	Table weight

Equation 4.2.11.4: Rack Pinion torque calculation

---

[14] For some examples of a friction coefficient, see Appendix F.

note

A rack pinion mechanism can be designed for two different modes: 1. Moving Rack (as documented so far) or 2. Moving Motor (plus Pinion). In the second operating mode the only inertia that applies is the pinion's inertia plus, in all consequence, the motor inertia (not the rotor inertia, but the entire motor and additional mounting components).

4.2.12   Rotary Table

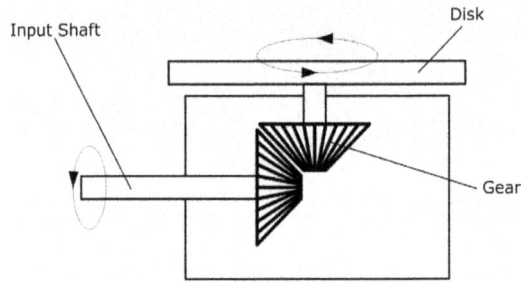

Picture 4.2.12.1: Rotary Table

A rotary table consists of a rotating disk and a gear, which may be a right-angle gear. The gear's data used to calculate the rotary table is the gear inertia, transmission ratio, friction coefficient and efficiency.

$$J_{Disk} = \frac{m \times r^2}{2} = \frac{W \times r^2}{2 \times g} = \frac{\pi \times L \times \rho \times r^4}{2 \times g}$$

g	Gravity Constant = 9.81 m-s^2 = 386 in-s^2
J$_{Disk}$	Disk Inertia
L	Disk Length/Thickness
m	Mass
ρ	Mass density of material
r	Radius
W	Weight

Equation 4.2.12.1: Disk Inertia

$$J_{total} = J_G + J_{L \rightarrow M}$$

$$J_{L \rightarrow M} = \left(\frac{1}{N_r} \right)^2 \times \frac{J_L}{e}$$

e	Mechanism efficiency
J_{total}	Total inertia including load
J_G	Gear inertia
$J_{L\text{->}M}$	Load inertia reflected to motor
J_L	Load inertia (includes disk and load)
N_r	Transmission ratio

Equation 4.2.12.2: Rotary table inertia calculation

$$T_{L \rightarrow M} = \frac{T_L}{N_r \times e}$$

e	Mechanism Efficiency
N_r	Transmission ratio
$T_{L\text{->}M}$	Load torque reflected to motor
T_L	Load torque

Equation 4.2.12.3: Rotary table load torque calculation

4.2.13 Center Driven Winder

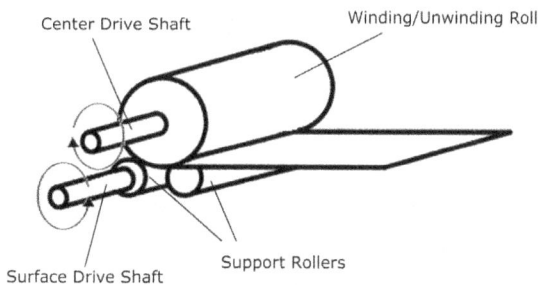

Picture 4.2.13.1: Winder

A center driven winder is a winder where the winding/unwinding roller is directly connected to the motor. Supporting rollers are optional, but if they are used it is mandatory to include them into the inertia calculation.

In principle, a winder contains of one winding/unwinding roller and a number of support rollers that can be calculated like a disk. The diameter of the winding/unwinding roller in relation to the support roller(s) determines the transmission ratio that will influence the inertia of the support roller(s) reflected to the motor.

note Naturally, the diameter of the winding/unwinding roller will increase/decrease during operation and thus change inertia and rotational speed. For the purpose of motor sizing, it is appropriate to apply the maximum diameter.

$$ J_{Disk} = \frac{m \times r^2}{2} = \frac{W \times r^2}{2 \times g} = \frac{\pi \times L \times \rho \times r^4}{2 \times g} $$

g	Gravity Constant = 9.81 m-s^2 = 386 in-s^2
J$_{Disk}$	Disk Inertia
L	Disk Length/Thickness
m	Mass
ρ	Mass density of material
r	Radius
W	Weight

Equation 4.2.13.1: Winding/Unwinding Roller Inertia

Equation 4.2.13.1 is also suitable to calculate the inertia of the support roller(s).

A winder is a load device, i.e. there is no other load attached to the winder. For that reason the transmission ratio is not significant for speed calculation, but it is needed to calculate the support roller(s) inertia that is reflected to the motor. The transmission rate of the winder is derived from the roller diameters.

note The following equations are based on one support roller. In all consequence, if more than one support roller is being used, they need to be calculated separately and their total inertia needs to be applied as support roller inertia.

$$N_r = \frac{D_{RL}}{D_{RM}}$$

N_r	Transmission ratio
D_{RL}	Diameter of support roller
D_{RM}	Diameter of winding/unwinding roller

Equation 4.2.13.2: Winder transmission ratio

The total inertia of the winder includes the winding/unwinding roller inertia and the inertia of the support roller(s) reflected to the motor.

$$J_{total} = J_{RM} + J_{RL \rightarrow M}$$

$$J_{RL \rightarrow M} = \left(\frac{1}{N_r}\right)^2 \times \frac{J_{RL}}{e}$$

e	Mechanism efficiency
g	Gravity Constant = 9.81 m-s^2 = 386 in-s^2
J_{total}	Total inertia
J_{RM}	Inertia of winding/unwinding roller
$J_{RL->M}$	Support roller(s) inertia reflected to motor
J_{RL}	Support roller(s) inertia
N_r	Transmission ratio

Equation 4.2.13.3: Winder inertia calculation

$$T_{L \to M} = \frac{T_L}{N_r \times e}$$

e	Mechanism Efficiency
N_r	Transmission ratio
$T_{L->M}$	Support roller(s) torque reflected to motor
T_L	Support roller(s) torque

Equation 4.2.13.4: Winder load torque calculation

One parameter that should not be neglected on a winder is the web tension, which should be added as a constant torque of the winding/unwinding roller.

4.2.14   Surface Driven Winder

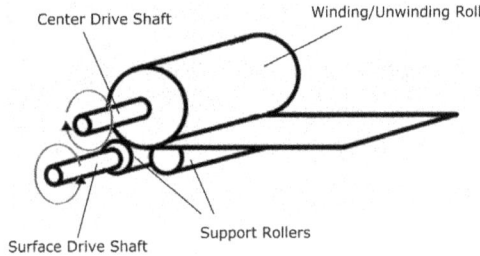

Picture 4.2.14.1: Winder

A surface driven winder is a winder where one support roller is directly connected to the motor. Further supporting rollers are optional, but if they are used it is mandatory to include them into the inertia calculation. In principle, a winder contains of one winding/unwinding roller and a number of support rollers that can be calculated like a disk. The diameter of the driven support roller in relation to the winding/unwinding roller determines the transmission ratio that will influence the inertia of the winding/unwinding reflected to the motor.

note

Naturally, the diameter of the winding/unwinding roller will increase/decrease during operation and thus change inertia and rotational speed. For the purpose of motor sizing, it is appropriate to apply the maximum diameter.

$$J_{Disk} = \frac{m \times r^2}{2} = \frac{W \times r^2}{2 \times g} = \frac{\pi \times L \times \rho \times r^4}{2 \times g}$$

g	Gravity Constant = 9.81 m-s^2 = 386 in-s^2
J$_{Disk}$	Disk Inertia
L	Disk Length/Thickness
m	Mass
ρ	Mass density of material
r	Radius
W	Weight

Equation 4.2.14.1: Winding/Unwinding Roller Inertia

Equation 4.2.14.1 is also suitable to calculate the inertia of the support roller(s).

A winder is a load device, i.e. there is no other load attached to the winder. For that reason the transmission ratio is not significant for speed calculation, but it is needed to calculate the support roller(s) inertia that is reflected to the motor. The transmission rate of the winder is derived from the roller diameters.

note

The following equations are based on one support roller. In all consequence, if more than two support roller is being used, they need to be calculated separately and their total inertia needs to be applied as support roller inertia.

$$N_r = \frac{D_{RL}}{D_{RM}}$$

N$_r$	Transmission ratio
D$_{RL}$	Diameter of winding/unwinding roller
D$_{RM}$	Diameter of driven support roller

Equation 4.2.14.2: Winder transmission ratio

The total inertia of the winder includes the driven support roller inertia and the inertia of the winding/unwinding roller reflected to the motor.

$$J_{total} = J_{RM} + J_{RL \to M}$$

$$J_{RL \to M} = \left(\frac{1}{N_r}\right)^2 \times \frac{J_{RL}}{e}$$

e	Mechanism efficiency
g	Gravity Constant = 9.81 m-s^2 = 386 in-s^2
J$_{total}$	Total inertia
J$_{RM}$	Inertia of winding/unwinding roller
J$_{RL->M}$	Winding/unwinding roller inertia reflected to motor
J$_{RL}$	Winding/unwinding roller inertia
N$_r$	Transmission ratio

Equation 4.2.14.3: Winder inertia calculation

$$T_{L \to M} = \frac{T_L}{N_r \times e}$$

e	Mechanism Efficiency
N$_r$	Transmission ratio
T$_{L->M}$	Winding/unwinding roller torque reflected to motor
T$_L$	Winding/unwinding roller torque

Equation 4.2.14.4: Winder load torque calculation

One parameter that should not be neglected on a winder is the web tension that should be added as a constant torque of the winding/unwinding roller.

Motor Sizing Programs

There are a number of motor sizing programs available in the market, most of them manufacturer-specific and some of them, such as VisualSizer by Copperhill Technologies, are manufacturer-independent. The manufacturer-specific versions naturally include only a limited number of motors, while the generic versions offer thousands of motors from a variety of manufacturers.

The motor selection process as described also explains the popularity of motor sizing programs. The process of recalculating the torque requirements for each individual motor/drive combination can be extremely time-consuming considering the vast amount of motors available in the industry. The goal of motor sizing is to find the optimum motor for the application and that can only be accomplished with sufficient choices available, i.e. with a great number of applicable motors.

Hear Ye! Hear Ye!

5.1 Motor Sizing Programs for Windows

Two very popular versions, not based on VisualSizer, are the Danaher Motioneering software or Alpha Gear's Cymex. VisualSizer in turn is the most used motor sizing software since it has been adapted by motor manufacturers such as Siemens Energy & Automation, Schneider Electric, Parker Hannifin, Oriental Motors, Baldor, Automation Intelligence, AutomationDirect.Com, GE Fanuc, Moog, and more. In the following, we will concentrate on

113

the VisualSizer motor sizing software.[15] The sizing process under VisualSizer is divided into four logical steps:

1. **Axis Design**

   Assign mechanical components and enter their data.

2. **Velocity Profile**

   Set up the duty cycle

3. **Motor Selection**

   Access the motor database

4. **Report Generator**

   Create an application report

In addition, VisualSizer provides a fifth screen to compare the application requirements with the motor performance.

5.1.1 Axis Design

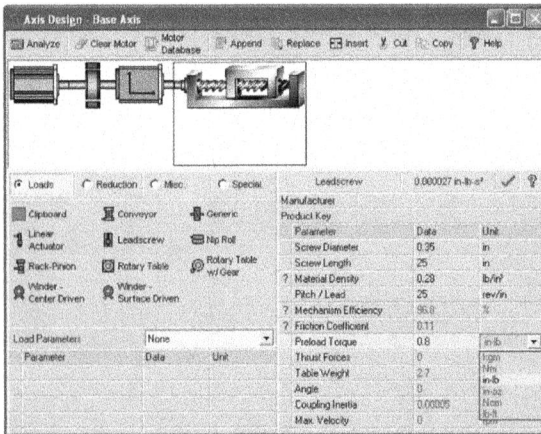

In this example, the user constructed a system including a motor, brake, right-angle servo reducer and a leadscrew. VisualSizer offers a great variety of mechanical devices that are separated by Loads, Reduction, Misc. and Special. The "Special" section contains of specific mechanical components such as Shimpo Servo Reducers which require a special selection process.

---

[15] For detailed information on VisualSizer-Professional log on to http://www.visualsizer.com.

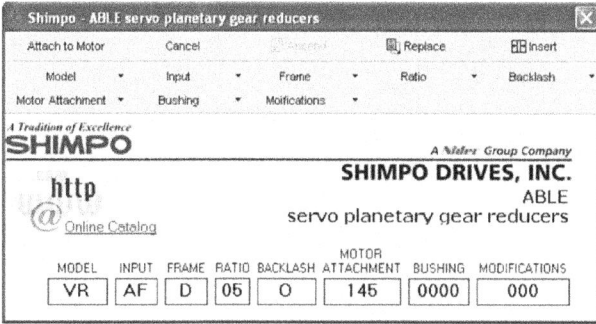

5.1.2 Velocity Profile

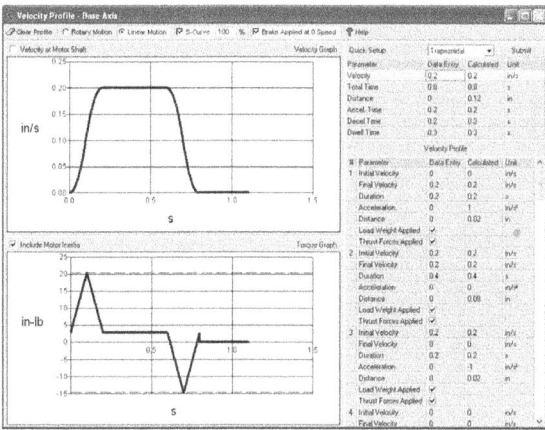

In this example, the user chose a trapezoidal velocity profile with jerk limitation (S-Curve) applied. The jerk limitation is set to 100%. The user has the choice to design up to 25 motion segments where each segment provides the option to apply the thrust forces and/or load weight.

The screen also shows both curves, velocity and torque.

5.1.3 Motor Selection

VisualSizer currently comes with a motor database of 5000+ motor/drive combinations, most of which also provide a performance profile (torque vs. speed).

The user also has a choice of preferred manufacturers and motor technologies. In addition, VisualSizer provides a database editing tool that allows the user to add and modify motor data.

5.1.4 Report Generator

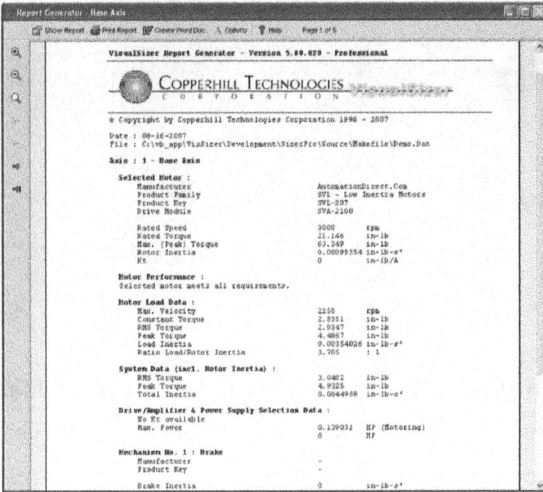

The report contains all user inputs, calculated parameters (for instance, for drive selection) and, of course, the selected motor. All graphs, velocity, torque and performance, are also included. In addition, the user can export the report into Microsoft Word and there he can modify the report to add further information. The template used for the Word file is also customizable.

5.1.5 Performance Curves

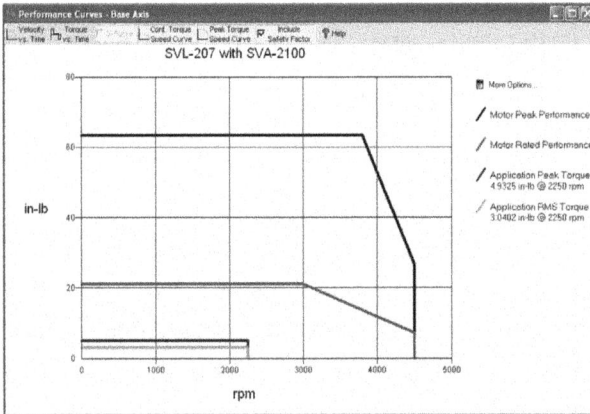

Last, but not least, VisualSizer provides an additional screen where the user can compare the application requirements with the motor performance. This graph is also included in the application report.

Appendix A - References

/1/
The Texonics Motion Cheat Sheet
Texonics, Inc.

/2/
The Smart Motion Cheat Sheet
Created by Brad Grant, P.E.
MSI Technologies, Inc., 1994

/3/
Optimizing motion-control-system design
By Andrew Wilson, Editor
Vision Systems Design, July 2005

/4/
Motor Sizing Made Easy
By Craig Ludwick
Motion Control, January/February 2001

/5/
Motor sizing made simple
By Walt Dryburg
Machine Design, January 20, 2005

/6/
Basic Physics for Drive System Engineers
By Peter H. Warner
Machine Design, January 20, 2005

/7/
Motor Sizing - Calculating Speed, Inertia & Torque Requirements for Electrical Motors
By Wilfried Voss
http://www.copperhilltech.com

/8/
Energy-Efficient Motors Deliver Savings
By Frank Bartos
Control Engineering, July 1, 2005

/9/
Efficient Motors Can Ease Energy Crunch
By Frank Bartos
Control Engineering, May 1, 2001

/10/
Efficiency to the Masses — of Electric Motors, that Is
By Frank Bartos
Control Engineering, July 1, 1998

/11/
Motor Guide
http://www.pge.com/biz/rebates/express_efficiency/useful_info/motor_guide.html

/12/
Technical Reference: Sizing Example
Oriental Motor General Catalog 2003/2004

/13/
Engineering Information
Siemens General Motion Control Catalog Part 1 – 1999

/14/
Motor Planning Kit
Version 2.0
Developed by the Motor Decisions Matter campaign

/15/
Parker-Trilogy Linear Motors
Motor Sizing Application Note
By Jack Marsh

/16/
Choosing A Motor
By Dr. J.C. Compter
Professor of Engineering at Eindhover University

/17/
Proper Servo Sizing Produces Savings
By Wilfried Voss
Design Product News, May 2006

/18/
Sizing servo motors
By Carl Vangsness
IC&S Special Report, November 1995

/19/
Maybe you don't want to match inertias
By Frank Arnold
Motion Control – July/August 1998

/20/
Direct Process Control Using n-Dimensional NURBS Curves
By Robert M. Cheatham, W. Edward Red and C. Greg Jensen
Brigham Young University
Computer-Aided Design & Applications, Vol. 2 No. 6, 2005, pp 825-834

/21/
The Myth of Inertia Matching
By Nick Repanich
California State University, Chico, 1992

/22/
Don't neglect thermal concerns when specifying motor needs
By John Mazurkiewicz
EDN, September 20, 1979

/23/
Optimize Motor Control by Matching Motor Types to Applications
By Chuck Lewin, Performance Motion Devices
RTC Magazine, March 2007

/24/
Energy Efficient Motion
By Mark T. Hoske
Control Engineering, July 2007

/25/
Sizing Up the Field
Design News, September 2006

Appendix B – Web Site References

/1/
http://www.copperhilltech.com

/2/
http://www.visualsizer.com

/3/
http://www.motorsmatter.org

/4/
http://zone.ni.com/devzone/cda/tut/p/id/3367
National Instruments Tutorial
Fundamentals of Motion Control
February 1, 2006

/5/
http://www.linengineering.com/dc
Lin Engineering Designer's Corner

/6/
http://www.orientalmotor.com/in_motion/
Oriental Motor
New Motion E-Newsletter

Appendix C – Symbols & Definitions

a, α	Acceleration
d	Deceleration
D	Diameter
E	Mechanism Efficiency
F_{fr}	Friction Forces
F_g	Gravity Forces
F_P	Push Pull Forces
g	Gravity Constant
J	Inertia
$J_{L->M}$	Load Inertia reflected to motor
N_r	Transmission Ratio
N_T	Number of Teeth
P	Pitch
t	Time
T	Torque
W	Weight
ω	Angular Velocity
θ	Rotation
π	"PI" = 3.141592654
μ	Coefficient of Friction
γ	Application Angle

Appendix D – Material Densities

Material	lb/in³	g/cm³
Aluminum	0.096	2.66
Brass	0.3	8.3
Bronze	0.295	8.17
Copper	0.322	8.91
Plastic	0.04	1.11
Steel	0.28	7.75
Hard Wood	0.029	0.8

Appendix E – Mechanism Efficiencies

Mechanism	Efficiency
Acme screw with brass nut	~0.35 – 0.65
Acme screw with plastic nut	~0.5 = 0.85
Ball screw	~0.85 – 0.95
Preloaded ball screw	~0.75 – 0.85
Spur or bevel gears	~0.9
Timing belts	~0.96 – 0.98
Chain & Sprocket	~0.95 – 0.98
Worm gears	~0.45 – 0.85

Appendix F – Friction Coefficients

Material	μ
Steel on steel	~0.58
Steel on steel (greased)	~0.15
Aluminum on steel	~0.45
Copper on steel	~0.3
Brass on steel	~0.35
Plastic on steel	~0.15 – 0.25

Mechanism	μ
Ball bushings	<0.001
Linear bearings	<0.001
Dove tail slides	~0.2
Gibb ways	~0.5

Index

www.ingramcontent.com/pod-product-compliance
Lightning Source LLC
Chambersburg PA
CBHW081543220326
41598CB00036B/6549